Palaeobiogeography of Marine Fossil Invertebrates

Palaeobiogeography of Marine Fossil Invertebrates
Concepts and Methods

Fabrizio Cecca
Universe Pierre et Marie Curie
(Paris VI), France

CRC Press
Taylor & Francis Group
Boca Raton London New York

CRC Press is an imprint of the
Taylor & Francis Group, an **informa** business
A TAYLOR & FRANCIS BOOK

CRC Press
Taylor & Francis Group
6000 Broken Sound Parkway NW, Suite 300
Boca Raton, FL 33487-2742

First issued in hardback 2019

ISBN 13: 978-0-415-28789-0 (hbk)

Cover: The palaeobiogeographical distribution of some selected ammonite species plotted on two palaeogeographic maps which illustrate the distribution of continental masses and oceans during the Late Jurassic. The upper and lower maps correspond to the Early Kimmeridgian (around 146 million years ago) and the Late Tithonian (approximately 135 million years ago) respectively.

British Library Cataloguing in Publication Data
A catalogue record for this book is available
from the British Library

Library of Congress Cataloging in Publication Data
A catalogue record has been requested

Typeset in Times NR by Graphicraft Limited

Contents

Preface

This book is based on material gathered from personal research and lectures delivered to undergraduate and postgraduate university students.

Experience of the vicissitudes of trying to gather together the necessary papers and special issues on the different aspects of biogeography suggested that access to biogeographical literature suffers from two basic problems:

1 only a fraction of the specialized journals are available in a single library
2 books dealing with palaeobiogeography generally either are multi-author volumes, gathering articles on particular subjects, or are limited to aspects of biogeography that provide only a partial treatment of general concepts and methods.

Palaeobiogeography is generally treated in articles for specialists who already have a background in biogeography. This may discourage the larger audience of both students and confirmed palaeontologists who have yet to deal with biogeographical issues. There seems to be no comprehensive review of the important theoretical concepts or discussion of the common analytical methods applied.

This book does not set out to provide historical descriptions of biotic provinces or biogeographically-based palaeogeographic reconstructions. Excellent examples of these are already available in Middlemiss *et al.* (1971), Hallam (1973, 1994), McKerrow and Scotese, (1990) and Briggs (1995).

As much of the experience on which this book is based has been on ammonites and Mesozoic pelagic biotas, I have been conscious of the potential for a 'lack of balance'. Numerous case histories based on other fossil groups and geological periods are therefore included.

Why 'palaeobiogeography of marine fossil invertebrates'?

Palaeobiogeography involves all fossil organisms, plants and animals of the Phanerozoic Eon, including both extinct forms and living representatives. A textbook based on such a large database is far beyond the aims of the present one.

The focus on marine organisms reflects my personal experience. However, fossils of marine invertebrates are quantitatively the most represented in the fossil record, and this therefore offers greater access to marine environments of the past. Thus, there is a reciprocal relationship between neobiogeography and palaeobiogeography, the former having almost complete access to continental environments while the latter is largely restricted to marine environments.

Van der Spoel (1983) rightly stressed that the biogeography of marine plankton corresponds to 'the biogeography of 70% of the earth surface and less than 5% of biogeographical literature (and probably also of the knowledge on biogeography)' (p. 291) and that biogeographical theories need isolated areas with barriers affecting dispersal and speciation, whereas in the open ocean 'any organism can move from any point in the sea to any other locality' (p. 292). This is why Chapter 3, devoted to classical theoretical models of biogeography, summarizes concepts mostly defined and developed by neobiogeographers working on terrestrial organisms but rarely used in palaeobiogeography, particularly where this relates to marine organisms.

The characteristics of the larval stages of marine invertebrates are vitally important to the understanding of most biogeographical patterns. In fact, marine invertebrates are subject to a variety of larval dispersion processes, some of which allow journeys of thousands of kilometres (depending on both the type of larval development and the velocities of the currents responsible) before metamorphosis and final settlement. The biogeographical implications of such processes are discussed at length in Chapter 6.

For these reasons, it is difficult to conceive a textbook that could encompass all aspects of the palaeobiogeography of both terrestrial and marine biotas. Although the focus here is on marine organisms, the reader will find in this volume the most important general concepts of biogeography which are also valid for terrestrial biotas.

Is palaeobiogeography a pure geological or a palaeobiological discipline?

Palaeobiogeography is, strictly, a distinct part of biogeography, as will be discussed later. It is one of the most important disciplines of the natural sciences because of the implications it has for evolution and its potential applications in the reconstruction of past geographical scenarios, both global and regional. It represents the interface between earth and life sciences and it is difficult to consider it either as a pure palaeobiological or geological discipline.

Attempts to explain palaeobiogeographical patterns lead to the generation of hypotheses about continental drifting, past migration routes (land bridges or seaways), palaeobiodiversity gradients and geographical barriers, in addition to hypotheses on past climatic and oceanographic conditions.

Although essentially exploited after Wegener's continental drift hypothesis was vindicated on the basis of geophysical evidence, the palaeontological (palaeobiogeographical) evidence of plate tectonics gave a new impetus to palaeobiogeography, both in the explanation of distributions and in its application to geotectonic problems. However, this sometimes entails circular argument when the same fossil data are used to infer biological processes and palaeogeographic reconstructions. Geological processes act to model the Earth and to change geographical scenarios. The closing of an ocean forms a barrier for marine organisms but may result in the creation of a route for terrestrial organisms. Therefore, as Hallam (1994) indicated, the mutual interaction between geology and palaeontology allows us to compare the results obtained using different fossil groups with geological data and to evaluate whether or not they are concordant. A hypothesis formulated to explain patterns inferred from one data set is accepted with greater confidence 'if it also accounts for patterns inferred from completely unrelated data sets' (Hallam, 1994, p. 6).

Is it possible to investigate palaeobiogeographical processes and yet ignore palaeobiology? It is hard to find a correct answer to this question, but in practice palaeobiogeography always needs information about the biological characteristics of fossil organisms and this is derived from comparison with modern closely related equivalents. I have tried to show this in some chapters. The close relationships between palaeobiogeography and biological evolution are inherent in models of geographical speciation that obviously imply a spatial dimension. Thus, in this sense, there is no clear boundary between biology and biogeography.

Acknowledgements

I could not have realized this book without the immense patience of Amina, my wife: I only can say that now we will not forgo any more holidays. But our holidays would be still very far in the future without considerable help from a number of colleagues and friends. These are, in alphabetical order: Martin Aberhan (Museum für Naturkunde, Berlin), Colin Braithwaite (The University of Glasgow), Neale Monks (The Natural History Museum, London), Nicol Morton (The Birkbeck College, London), Peter Rawson (The University College, London) and Brian Rosen (The Natural History Museum, London). They all kindly gave me constructive criticisms, helpful comments, and suggestions and/or linguistic assistance on the former versions of the chapters of this book: thank you very much! I also would like to thank Bruno Vrielynck (CNRS, University of Paris VI) for supplying the palaeogeographical reconstructions used by Gilles Simonpiétri (Université de Provence) to draft both the cover of the book and Fig. 8.10. The Centre de Sédimentologie et Paléontologie of the Université de Provence (Marseille) also provided help. I am grateful to a number of Gordon & Breach and Taylor & Francis employees who worked with me during the initial phases of this project and who supervised the practical realization of the book.

I would like to recall here that my interest in biogeography was strongly supported and stimulated by Eric Fourcade during the two years we worked together on the maps of the 'Programme Tethys'. I deeply regret that he cannot read this book. Thanks Eric.

The majority of the illustrations have been copied from previously published books and scientific articles. Permission to reproduce these has been obtained from the authors, journals and publishers concerned. I would like to thank: Academic Press LTD (London), Backhuys Publishers (Leiden), Blackwell Science, *Bulletin of Marine Science* (Miami), Bureau des Recherches Géologiques et Minières (Orléans), Cambridge

University Press, Casa Editrice Ambrosiana (Milano), Elsevier Science (Amsterdam), Geobios (Lyon), Lethaia (Oslo), Linnean Society of London, *Nature* (London), Oxford University Press, Oregon State University Press, SEPM – Society for Sedimentary Geology (Tulsa), Société Géologique de France (Paris), *Systematic Biology*, The Geological Society of America, The Geological Society of London, The Royal Society (London), The US Paleontological Society and Paleobiology, The University of Chicago Press, A. J. Boucot, L. R. M. Cocks, R. A. Fortey, C. Mellish, B. Rosen, A. Smith, D. Jablonski, J. W. Valentine, K. Roy, C. W. Stock, M. Zunino.

Fabrizio Cecca
Marseille, May 2001
(When writing this book the author was affiliated
to the Université de Provence (Marseille))

1 Introduction

1.1 Processes of speciation and biogeography

It has been recognized since the pioneering work of Darwin that biological evolution has a biogeographical component. The biological concept of the species (Mayr, 1940), which strictly applies only to organisms that reproduce sexually, suggests that species are separated by mechanisms that form barriers to gene flow and isolate them. During evolution, new species arise while others disappear and the biogeographical component of evolution acts to separate them physically. Geographical barriers may form reproductive barriers and prevent the flow of genes between emerging species or between populations of the same species.

Although there is general agreement on the reality of geographical speciation, its importance in producing the number of species now living is a matter of discussion and several non-geographical processes of speciation have been described. Rosenzweig (1995) suggested that three basic types of speciation can be recognized:

1 geographical speciation
2 polyploidy
3 competitive speciation.

Geographical speciation is more widely known as **allopatric speciation** and competitive speciation is an expanded version of **sympatric speciation** (a sympatric non-polyploid speciation).

Polyploidy is the most rapid and geographical speciation the slowest of processes of speciation. The latter may take thousands or even hundreds of thousands of years, whereas competitive speciation may only require between ten and several hundred generations, as shown by laboratory experiments on *Drosophila*. Polyploidy is independent of both geography and ecology, whereas competitive speciation depends on ecological opportunities. Geographical speciation also depends on ecology because long-term environmental differences between areas containing an original biota may lead to the creation of real isolates and then to the development of endemic taxa.

Competitive (sympatric) speciation requires disruptive selection, the separation of a unimodal phenotypic distribution into a bimodal one. This occurs because the mode becomes unfit for an environmental system favouring two new and substantially different phenotypes. The process involves the evolution of reproductive isolation followed by the elimination of all intermediate forms. This stable polymorphism occurs in a heterogeneous environment. Rosenzweig (1995, p. 98) defined **competitive speciation** as 'the expansion of a species from a single ecological opportunity to an unexploited ecological opportunity followed by that species' sympatric breakup into two daughters, one using the original opportunity, the other the newly exploited one'. In this model the expansion of the species happens as a result of a lack of competition for the use of a previously unexploited ecological opportunity. However, the speciation that follows reflects an excessive competitive pressure on intermediate phenotypes. Rosenzweig considers that while competitive speciation is usually rare, it may be relatively rapid as far as geological time is concerned. This model may therefore provide one of the mechanisms of so-called quantum speciation corresponding to short-term, relatively uncommon episodes of ecological opportunity.

Partly because of theoretical difficulties, competitive speciation has been poorly represented in the studies of most evolutionists. However, recently proposed models of competitive (sympatric) speciation by Tregenza and Butlin (1999), Dieckmann and Doebeli (1999), and Kondrashov and Kondrashov (1999) provide a firm theoretical basis for the sympatric origin of numerous species.

Although geographical speciation is the dominant process (Rosenzweig, 1995), it is impossible to dismiss the possibility that speciation can occur without geographical isolation. More recently, Norris (2000) has observed that sympatric and parapatric models of speciation are more satisfactory for explaining the high diversity in pelagic taxa.

1.1.1 The geographical dimension in speciation

The emphasis here is on the evolutionary outcomes of changes in the geographic ranges of ancient species.

The existence of geographical variation in spatially separated populations within a single species led to the development of the theory of **allopatric speciation**, that is, speciation due to geographical separation. The prerequisite to such speciation is a **geographical isolation**, produced by strictly extrinsic factors, the appearance of seaways, rivers, mountains, etc., dividing a former single gene pool into two. Two populations of the same species become geographically separated, either by the development of some physical barrier or by the elimination of the intervening populations, diverging from each other as a result of the interruption of the gene flow over a period of time. If the barrier is eliminated, the two populations may again interbreed if their genetic differentiation has been small, but otherwise speciation has occurred and two new species have been created (Mayr, 1963).

The term **peripatric speciation** (or **founder effect speciation**) was proposed by Mayr (1942) to describe the evolutionary role of small peripheral populations that become separated from the main population and colonize areas not previously occupied by it. Such peripheral populations are called founder populations or **propagules**. They may contain only a fraction of the genetic variation of the original population but are nevertheless sufficient to establish a new breeding population. Divergence from the main population, and ultimately speciation, occurs rapidly because the founder population is genetically distinct from the parent species in containing only a reduced portion of the original genetic variation. However, many biologists do not accept that founder effects are important in speciation in newly colonized areas, believing that the process should be driven by natural selection rather than genetic drift. The rapid multiplication of species to occupy new environments is called **adaptive radiation**.

The prerequisite for speciation is the division of the gene pool by a barrier such that speciation appears to be a 'passive' phenomenon. Biogeographers call this **vicariance**. However, an 'active' phenomenon, **dispersal**, can also explain allopatric and parapatric speciation. A stock of the ancestral population can cross a barrier and colonize a new area. The subsequent reduction of the gene flow will produce two allopatric species. **Parapatric speciation** implies only partial geographical separation between two contiguous speciating populations and does not require the creation of a barrier to stop the gene flow. This model assumes that a small peripheral population comes under a strong selective pressure different from that directed at the main population. Garagna *et al.* (1997), on the basis of the chromosomal characteristics of a population of mice living in a wood near Seveso (Northern Italy), have demonstrated that a process of speciation (by chromosome rearrangements) occurred with an evolutionary 'jump' produced in no more than 20 years. The origin of this new species has been traced from a parapatric nearby population and has been related to time of massive material movements during expansion and contraction of an urban area.

To conclude this review, it appears that biological evolution is connected with changes in the areal distributions of organisms. Biogeography is therefore relevant to the study of evolution.

1.2 Biogeography: neo- and palaeobiogeography

Biogeography deals with the geographical distribution of both living and fossil organisms and with the processes that brought about this distribution. Biogeographical distributions are a reflection of processes acting at a variety of temporal and spatial scales.

Rosen (1992, 1995) suggested that the terms **neobiogeography** and **palaeobiogeography** can be used to describe 'special cases' of biogeography because they define the endpoints of a continuum. We can refer to this entire continuum as biogeography, despite the fact that some authors have confined the term to indicating only the study of the distributions of living organisms. Although philosophical and conceptual differences are not profound, there are fundamental methodological contrasts between neo- and palaeobiogeography that derive from the practical problems encountered in working with fossil data.

Neobiogeography studies the distributions of organisms whose geographical limits and environmental conditions can be directly observed on the surface of the modern Earth. Distributions may be quantified on the basis of standard subdivisions of a grid referred to topographical maps, for example, the number of individuals of a given species on a standard geographical surface. Crovello (1981) proposed the terms **operational geographical units** and **operational palaeogeographical units** (**OGUs** or **OPGUs**) to describe these subdivisions.

By contrast, palaeobiogeographers are forced to work with assumptions about ancient geographical and environmental conditions. In particular, they must reconstruct palaeogeographies and palaeoenvironments from indirect evidence, before studying the fossil distributions. Thus, quantifications of species diversity are not equivalent to those of neobiogeographers. Where neobiogeographers refer to standard surfaces palaeontologists work with hypothetical geographies.

A possible boundary, if a boundary is really needed, between neobiogeography and palaeobiogeography might be taken at the significant modification of Earth's landscape, represented by the last glacial expansion.

The taxonomic methods of the two groups are obviously different. Biologists are able to test numerous cases of the real specific position of closely related organisms. By contrast, the incompleteness of the fossil record constrains palaeontologists, who can only evaluate the morphological characters of the hard parts of the organisms. They must also assume that the population that they see truly reflects the full diversity of the original population. Recent researches in molecular biology have provided a new and important tool to neobiogeography. For example, the recent study by Cavalli-Sforza *et al.* (1997) of human genes and their biogeographical implications radically changes our views of morphological variability and the relationship between the human genome and the phenotypic expression of characters. Neobiogeography is more able to investigate ecological and environmental factors in biogeographical processes than palaeobiogeography, but it can only study short-term processes acting on a biological time-scale.

Palaeobiogeography, however, has the advantage of direct access to the temporal dimension, thus focusing on changes in distributional patterns through time and avoiding the assumptions on evolutionary pathways used by neobiogeographers. The distributional patterns are typically referred to a stratigraphical time-scale. This difference has been succinctly summarized by Rosen (1995, p. 300): 'In general, palaeobiogeography is stronger on patterns than process'. Geologists are familiar with the use of palaeobiogeographical data for palaeogeographical reconstructions. However, the biogeographical distributions of living organisms are also used to infer past geographical scenarios through cladistic methods and panbiogeography, although the use of fossils is more common. Geological problems are greater in neobiogeography when long-term processes such as distributional change and origination are considered.

1.3 The aims of palaeobiogeography

This subject has generated significant debate in the biogeographical literature.

Patterson (1983), following Wallace (1876, 1880), summarized biogeographical aims as of two kinds:

1. to use distributions as a guide to Earth history
2. to explain distributions through application of theories of Earth history and evolution.

The first aim is investigative and seeks to identify patterns, whereas the second is explanatory and seeks to explain patterns by process. It is evident that circular reasoning is involved in these conflicting aims and it is important to avoid basing biogeographical conclusions solely on biological and palaeontological evidence, or geological conclusions solely on geological evidence.

In palaeobiogeography it is sometimes difficult to differentiate biologically-derived and physically-derived hypotheses, particularly where the same fossil data have been used to infer biological processes and assist in palaeogeographical reconstructions (Rosen, 1992). Palaeobiogeography has numerous implications for strictly geological subjects, but genuine palaeobiogeographical results can be obtained. Conversely, biogeographical distributions can be used for geological purposes such as plate reconstructions.

The reconciliation of biogeographical data with palaeogeographical reconstructions on the basis of plate tectonics theory has been severely criticized. Patterson (1983, p. 5) stated that 'this approach can be no more profitable than was fitting biogeography to the fixed continents axiom of Darwin and Wallace'. While this is a little overstated, the reasons for it are obvious: the further back in time one goes, the less informative modern distributions are likely to be. In this context Hallam's (1994, p. 34) comment is important:

> The evidence for plate tectonics is immense, and the quality of the science has improved immeasurably since earlier in the century when continental drift was generally dismissed out of hand . . . It behoves biogeographers to respect this fundamental change, and the solidity of the general outlines of geological history that have since been worked out, just as palaeontologists must make sure that their own hypotheses are biologically plausible.

The first of Patterson's aims should correspond to the principal applications of palaeobiogeography, but he refers to the use of distributions of living organisms, more than fossils, to infer Earth history. However, older taxa, living when specific events took place, became extinct and were replaced by new taxa reflecting more recent geographical patterns. Distributional patterns fade with geological age, making historical events derived from analyses of extant biotas of little use in the reconstruction of Earth history. Palaeobiogeography interacts with geology by an iterative process of reciprocal illumination to a much greater extent than neobiogeography.

The second aim, to explain distributions, has been accepted by most biogeographers, but Patterson (1983) noted that it involves or invokes processes and therefore carries the risk that it may not be possible to separate the factors affecting pattern and process. In support of his view Patterson quoted Eldredge's (1981) view: 'explanation of the pattern in terms of ideas about process . . . just so stories . . . usually criticized as untestable inductive narratives'. According to Ball (1976, p. 411), who first introduced the term in biogeography, **narratives** are inductive explanations specially invented to match particular requirements and have little validity without independent data. Narratives are not aimed at the discovery or confirmation of a law, they do not establish new empirical facts and make no positive predictions about what may be found in future investigations. Rosen (1988a) pointed out that biogeography has often been criticized as lacking scientific rigour and a more analytical, hypothesis-testing approach has been advocated. However, it is impossible to adhere strictly to hypothetico-deductive methods alone, and elements of an inductive approach are indispensable.

The aims of palaeobiogeography can be better defined once a conceptual framework has been outlined, particularly the investigative pathways or rationales, and the limits and constraints (see Chapter 4). Rosen (1988a) distinguished:

1 'pure' biogeography, whose aims include explanations of distributions, and investigation of the processes concerned
2 'applied' biogeography, whose aims reflect concerns with ecology, evolutionary biology and geology.

The applications here can be matched with the first of Patterson's aims, while the 'pure' biogeography includes the investigation of processes that in Patterson's second aim are supposed to explain the pattern. But these authors seem to use the term 'process' in different ways. Patterson reduces it to ideas of dispersal and vicariance, whereas Rosen considers three major categories of processes (maintenance, distributional change and origination) acting at different temporal scales.

Due to the nature of the fossil record, 'pure biogeography', which focuses on biogeographical patterns and processes, is more closely allied to neobiogeography than to palaeobiogeography. In fact palaeobiogeographical data generally refer to long-term time intervals (compared to the time-scales of biological processes), even if the best available dating methods permit the distinction of the smallest intervals. This conceptual framework entails the application of a long-term (geological) database that does not allow the study of either maintenance processes or any other processes acting on a time-scale shorter than the time interval identified. The description of palaeobiogeographical distributions suffers from the incompleteness of the geological record. Palaeobiogeographical distributions cannot be easily described with quantitative techniques because palaeontologists deal with samples of the original populations whose statistical representativeness is unknown (Henderson and Heron, 1977).

For this reason it is convenient to split Rosen's aim 1 into two distinct objectives: **descriptive** and **causal**. However, it is possible to recognize a distinct 'applied palaeobiogeography', which can be considered as 'classical'. This is aimed at the reconstruction of evolutionary pathways (applied 'biological' palaeobiogeography) and past geological conditions, including palaeoclimates and palaeoceanography (applied 'geological' palaeobiogeography), and numerous case histories illustrate the potential applications.

1.4 The different biogeographies

1.4.1 Descriptive biogeography

Biogeographical distributions need to be recognized and described. Historically this has been done through the definition of biogeographic units such as regions and provinces (see Chapter 4). This descriptive approach, whether qualitative or quantitative (phenetic), is process-neutral and distinct from the causal aspects of biogeography. This does not, however, mean that the descriptive or phenetic approach is a lesser child of biogeography; on the contrary, it is a fundamental preliminary step in each biogeographic analysis where the aim is the description of distribution.

Both qualitative and quantitative approaches have been used to describe distributions, the former being considered subjective, or more subjective than the latter. Because of this high subjectivity and lack of rigour, the descriptive approach has been criticized (Patterson, 1983) and the definitions of provinces questioned (Rosen, 1984).

Usually the description of biogeographical distributions begins with mapping the distribution of the studied organisms. This is done using geographical maps of modern Earth before those indicating palaeogeography. It is important to note that the quantitative approach is more common in neobiogeography than in palaeobiogeography, because of the limits of the fossil record, taxonomical problems, and the hypothetical nature of the palaeogeography. However, the final result is the definition of biogeographic units such as provinces and regions, whose qualitative or quantitative definition depends on the goals of the study or on the quality of the fossil record.

It must be stressed that if the use of biotic distributions is aimed at palaeogeographical reconstruction, the plotting of fossil distributions on palaeogeographical maps may entail circular reasoning reflecting the hypothetical character of palaeogeography. A more rigorous approach is to keep biotic distribution patterns separate from reconstructions based on other evidence. Except for the initial compilation of data, it is better to use simple graphical methods such as area cladograms, area dendrograms, and area clusters to depict distributional patterns (Rosen, 1988b).

1.4.2 Causal biogeography

Biogeographical studies concerned with investigating the causes of distributions fall under this heading. Here reside the theoretical controversies between the principal biogeographic theories on the one hand and a more philosophical approach (inductive versus deductive reasoning) on the other. Two different approaches to causal biogeography are often distinguished, perhaps too rigidly (Rosen, 1988a). These are, respectively, ecological biogeography and historical biogeography. They can be differentiated on the basis of the temporal scale on which the various factors that determine the biogeographical distributions operate. Rosen (1992, 1995) suggested that what is commonly studied as 'ecological biogeography' corresponds, more or less, to what he refers to as 'maintenance processes', whereas distributional change and origination together broadly correspond to 'historical biogeography'. These rigid subdivisions are difficult to recognize in biogeographical studies that often deal with aspects that cannot be separated into well-defined categories. For example, the long-term persistence of environmental factors such as climate, temperature, and salinity may have historical consequences in determining the emergence of new biotas. However, given the aims of this book, these distinctions will be briefly discussed below. Particular aspects of these two categories relevant to palaeobiogeography will be developed in Chapters 5 and 7.

1.4.2.1 Ecological biogeography

Ecological factors are studied with respect to their influence in controlling biogeographical distributions. They act in the short term, at a temporal scale that nearly coincides with that of the biological cycles. Ecological biogeography is therefore essentially part of neobiogeography. There is a spatial component in many ecological processes because both abiotic and long-term geological processes interact with short-term biotic processes to produce biogeographical distribution patterns. The scale of this spatial component may be used to separate 'pure' ecology from ecological biogeography. For example, ecological factors acting at regional or continental scales can be studied from a biogeographical perspective. The role of ecological factors in controlling palaeobiogeographical distributions is a matter of discussion. Hallam (1994) recalls that ecological biogeography is concerned with processes acting at short temporal and spatial scales. Because the temporal scale is that of the biological processes, it is obviously impossible to appreciate the importance of the different ecological factors in determining the palaeobiogeographical distributions. Nevertheless, the publication of MacArthur and Wilson's (1963) equilibrium theory of island biogeography (ETIB) provided a strong impetus to both ecologists and neobiogeographers and also an awareness of the geographical dimension of a number of processes traditionally studied by ecology. Therefore, ecological biogeography is particularly concerned with the study of the relations between biodiversity and surface, such as island biogeography or the geographical distribution of the modern biodiversity. In Chapter 7 the basic concepts of, and ideas about, relations between palaeobiodiversity and palaeobiogeography will be illustrated and discussed with respect to evolutionary implications.

1.4.2.2 Historical biogeography

Historical biogeography is the study of:

1 factors that have governed the history of the geographical distributions of organisms
2 historical explanations for the geographical distribution of organisms with respect to evolutionary and geological processes.

Palaeobiogeography (as opposed to biogeography as neobiogeography) has sometimes been taken to mean 'historical biogeography' (Patterson, 1981). However, these terms are not synonymous because historical biogeography also deals with the distributions of living organisms. The adjective 'historical' refers to the long-term time-scale of the processes on which biogeography focuses. These processes are responsible for the historical pathways of changes in the biogeographical distributions, intended in a dynamical, evolutionary, sense. Factors are essentially represented by changes in the Earth's surface: in the extent of seas on cratonic areas, with the creation or destruction of seaways, with plate movements and realignments, the creation of mountain ranges, and oceanographic and climatic changes.

Historical biogeographical patterns stress the steps of biological evolution in its spatial dimension. This certainly gives scope for studies of phyletic relationships relative to the evolution of their geographic distributions, for example cladistic biogeography, but also allows geology to exploit biogeographical data for palaeogeographic reconstructions (see Chapter 5).

1.4.3 Applied biogeography

The geographical distribution of organisms is used to infer the ecological role of factors such as temperature, water depth, primary productivity, or levels of radiant energy, to investigate whether diversity and extinctions are influenced by levels of provinciality, or to infer past tectonic events and palaeogeographical configurations.

The application of the spatial distribution of both living and fossil organisms to plate tectonics is certainly the most 'popular' image of biogeography among geo-scientists. Biogeographical data were among the first evidence presented in support of the then novel theory of continental drift. More recently, biogeographical constraints have played an important role in the palaeogeographical reconstructions of the Lower Palaeozoic (McKerrow and Scotese, 1990; Hallam, 1994). Only some biogeographical concepts and processes interface with Earth history and only some of these have any practical value in inferring past palaeogeographical configurations and associated tectonic events.

The biogeographical literature offers a variety of methods of inferring Earth history through analyses of living and fossil distributions (see Chapters 5 and 8). However, different constraints are introduced by the use of fossils for such purposes.

2 General biogeographical concepts

This chapter focuses on biogeographic patterns and processes. Processes must be distinguished from patterns (from which processes are inferred). Biogeographical processes may correspond to different concepts and ideas of processes believed to be important parts of the biogeographical system. They will be discussed in detail in the second part of this chapter.

2.1 The main distributional patterns

The distributional patterns are identified in the biogeographical vocabulary by numerous terms. It is important to define how these terms are commonly intended and how they are intended in this book.

2.1.1 Distributional area or range

The range or biogeographical distribution of a species corresponds to its occurrence censused at a significant number of OGUs (or OPGUs) which cluster in a more or less continuous portion of space. This definition also includes the concept of reproductive area, although this is hard to demonstrate in palaebiogeography. The limits of the distributional area generally correspond to physical discontinuities (geographical, climatic, oceanographic) but may also depend on biotic factors (interaction with other species, etc.). Boucot and Gray (1983) recognized two orders of biogeographic control: the first order refers to the latitudinal distribution of the organisms, as controlled by north–south temperature variations, solar energy and seasonality gradients; the second order refers to longitudinal assemblages that are reproductively isolated by physical barriers including the distinctive physical properties of water masses, and distance. The latter implicitly refers to the modern eastward declining gradient of diversity observed in the Pacific Ocean, although long larval transport may overcome the effect of distance as a reproductively isolating factor (see Chapter 6). In neobiogeographical studies, quantitative methods based on statistical treatments of coefficients of similarity are commonly used to define ranges. In palaeobiogeography their use is questionable and probabilistic methods are preferred (see Chapter 4). The definition of the boundaries of the distributional area may be difficult in some cases, particularly for organisms with a patchy spatial distribution such as reef corals. Thus, the total distributional area of a taxon, its reproductive range, must be taken into account and should coincide with the distributional area of the **metapopulation** (see Chapter 6), which comprises all of the subpopulations of a species living in scattered patches. This also depends both on the mode of life of the organisms (sessile, vagile, planktonic) and the characteristics of their larval transport. The idea of distributional area may therefore change when planktonic or benthic organisms are dealt with. The distribution of sessile organisms is constrained by the two dimensions of the bottom surface and their extent strongly depends on larval transport and recruitment. By contrast, the biogeographic range of planktonic organisms implies a third dimension, depth, because most species move upwards and downwards daily. The total distributional area of a taxon consists of a reproductive range, surrounded by a non-sterile and a sterile expatriation area (Van der Spoel, 1983). Zunino and Zullini (1995) defined the distributional area of a species as the portion of space where it occurs interacting with the ecosystem in a non-ephemeral way. They also observed that the distributional area is a historical entity formed by a series of subsequent steps. More recently Zunino (1999) proposed the concept and the term **nemophoront** (bearer of places) to designate each successive step of a distributional area during its evolution (Fig. 2.1). In this way we can define the distribution area of a species as the largest unitary succession of nemophoronts. In Fig. 2.1 changes in areal extent at any historical step do not have evolutionary effects because the ephemeral disjunctions either are too short to provoke reproductive isolation or form part of the distributional area of a metapopulation.

Chirat (2000) demonstrated that the cosmopolitan distribution of the Tertiary nautiloid genus *Aturia* results from exceptional post-mortem transport of empty shells. In similar cases the inference of the real distributional range of a fossil taxon may not be clear and the recorded distribution (cosmopolitan in the case of *Aturia*) should be considered a **pseudodistribution** (this term is mine).

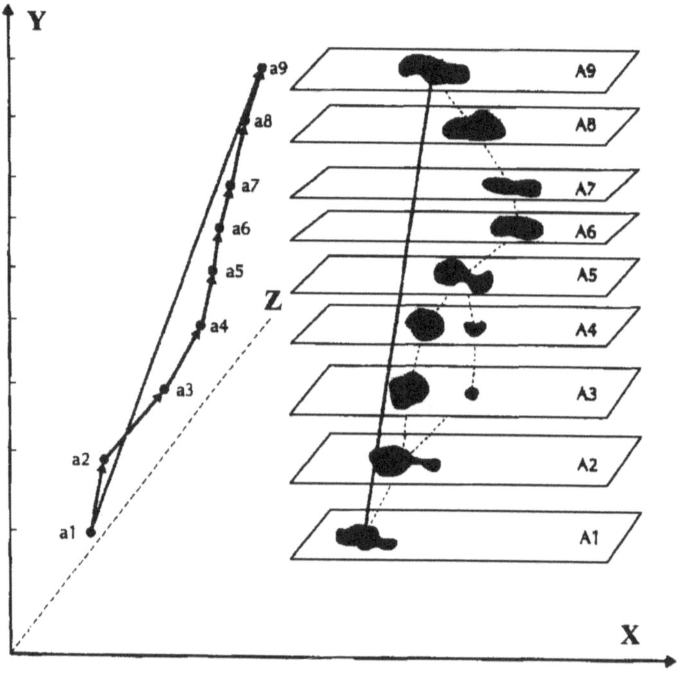

Figure 2.1 Steps of successive nemophoronts of the distributional area **A** of the species **a** along its anagenetic evolutionary path. X, longitude; Y, time; Z, latitude.

Source: modified from Zunino and Zullini (1995), with permission of the authors and Casa Editrice Ambrosiana.

2.1.1.1 Settlement, recruitment and colonization

Benthic ecologists define **settlement** as referring to the point at which larvae attach themselves, whereas the term **recruitment** refers to the stage when new members of the community become visible to be censused. Dispersion can lead to reproduction and final settlement in a new area, provided that the ecological conditions, particularly the temperature requirements for sexual maturation, are met. It has been demonstrated in modern reefal scleractinians (Hughes *et al.*, 1999) that variation in recruitment does not match variation in the abundance of adults. If the species recruited is able to retain its residence in the new area for a sufficient time-interval and eventually give rise to offspring, **colonization** has occurred in that area. Contrary to expectations, the chances of final settlement do not increase with reduced

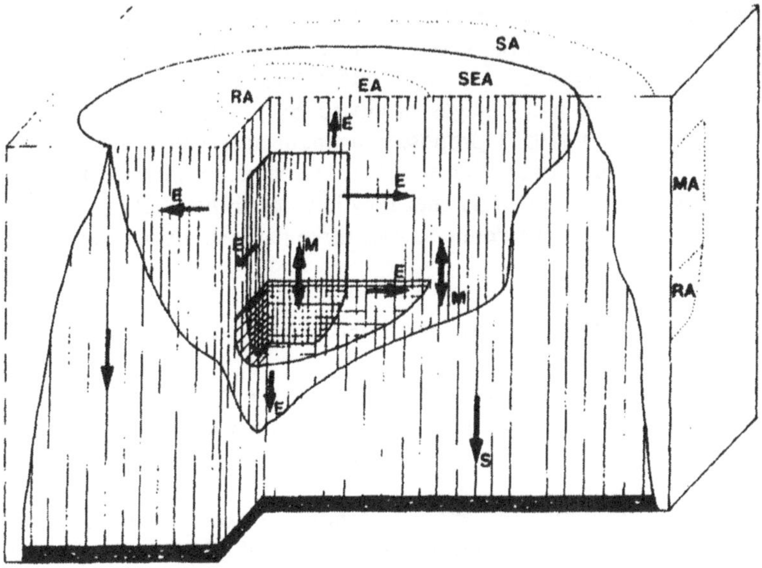

Figure 2.2 Theoretical model showing the relationship between the range of planktonic taxa and the reproductive area. E = the direction of expatriation; EA = the non-sterile expatriation area; M = migratory movements; MA = the migratory area; RA = the reproductive area; S = sedimentation; SA = sediment range; SEA = the sterile expatriation area.

Source: van der Spoel, 1983. Reproduced with permission of Academic Press.

distances of dispersal. Larval transport can produce repeated arrivals of larvae in a given area that reach the adult stage but do not achieve sexual maturity. Such non-reproductive '**pseudopopulations**', also called 'expatriate non-reproducing populations' by Scheltema (1971), are known to occur outside the area of the interbreeding population, maintained by means of continuous dispersal (Lutz and Jablonski, 1982; Bouchet and Taviani, 1992). This phenomenon has also been described in planktonic organisms (Fig. 2.2) and is referred to as **expatriation** by van der Spoel (1983). It explains the anomalous occurrences of tropical species off California and Peru during El Niño periods.

2.1.1.2 Residence

Resident populations are those that live and mate in an area. These two criteria are important in biogeography in defining the boundaries of the distributional area of a species. However, breeding can also occur in places where the species migrate ('migration' is here used in its biological sense; see Chapter 4). In palaeontology, particularly for nectonic organisms, it can be difficult to separate true resident populations from assemblages produced by post-mortem processes. Fernandez-Lopez (1991) proposed distinguishing **demic** and **ademic** species, the former indicating that fossil remains are inferred to be in the living area and the latter that they have been distributed from the living area. Among the demic species three basic situations have been distinguished:

- **eudemic**, if the fossil species is inferred to be in its life area and to have bred there
- **miodemic**, if it is inferred to have lived there without breeding
- **parademic**, if the area was occasionally reached by passive dispersion during life.

In the case of ammonites, eudemic conditions can be inferred on the basis of the occurrence of both macroconch and microconch morphotypes, interpreted by ammonite specialists as representing males and females of the same species. Fernandez-Lopez also considered the presence of juveniles. The term **eudemic** was introduced by Callomon (1985, p. 63) to describe species and genera living and breeding in a given area. However, when first proposed, it was used in an interpretative sense because Callomon (1985) referred to the evolution of successive species forming a lineage, making explicit reference to the 'centre of evolution' of the taxon, that is, the area where the fossil species is considered to have been eudemic.

2.1.2 Endemism

This concept was introduced by the French botanist de Candolle (1820). It indicates the exclusive distribution of a taxon (the endemic taxon) within a specific area. The resulting pattern of distribution is termed the endemic distribution. The notion of endemism is contained in the concept of 'provincialism' but this is not a synonym because biogeographic provinces, and hence provincialism, can be defined even in periods when intense faunal exchanges are taking place. Endemism is a relative concept: a taxon occurring in a given region and nowhere else is endemic to that region. However, when its distribution within the region is observed, it may be found to be endemic only to some portion of the region. Fig. 2.3 illustrates this concept.

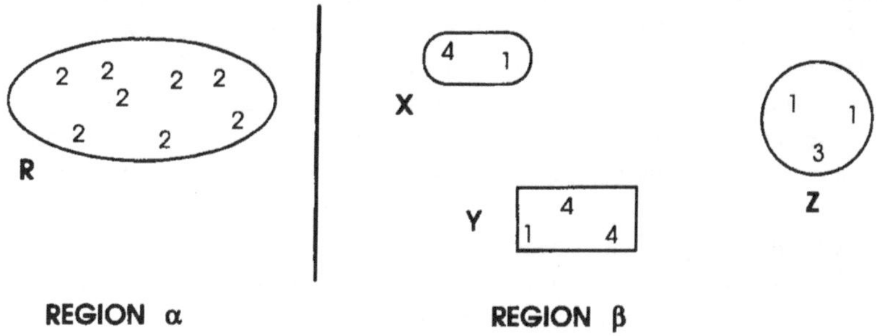

Figure 2.3 The distribution of four hypothetical taxa 1–4 in four hypothetical areas R, X, Y, Z. The solid line delimits regions α and β. Taxon 3 is endemic in a single locality of area Z (= autoendemic). Taxon 1 is endemic to region β and occurs in areas X, Y and Z. Taxon 2 is endemic to area R where it is widespread, but is absent elsewhere in region β. Taxon 4 is endemic to areas X and Y but can be considered endemic to region β when the biotic distributions of both regions are analysed.

The recognition of endemism from the distributional data of fossils forms the basis for all palaeobiogeographical analyses (Smith, 1988) and these must be based upon prior, sound systematic studies. The taxonomic level of the recognized endemic distributions must be specified (e.g. genus-endemic, family-endemic), as well as the time-interval during which the endemic distribution is observed (e.g. during the Late Jurassic, during the Kimmeridgian, or during the Cavouri Zone).

2.1.3 Disjunct distribution

Taxonomically related taxa living in areas largely separated by barriers show a disjunct distribution, or disjunct endemism. The disjunct distribution of a taxon is described by its presence in some localities at the range endpoints and by its absence in intervening areas. Palaeontologists should not therefore disregard the possibility that the absence of a taxon may be due to any or all of:

1 sampling failure
2 stratigraphical gaps
3 the lack of suitable environments for the adult organism to settle.

One of the major disjunct distribution patterns on the present surface of the Earth is **bipolarity**, the presence of phyletically related groups of organisms in the high latitudes of both hemispheres separated by a distinct equatorial gap. This distribution is also called **amphitropical** or **antitropical**. Among marine organisms, this pattern is shown by both fossil and extant molluscs and fishes (Crame, 1992, 1993).

The possibility that endemicity at any one site is a reflection of how much sampling has been done rather than an index of biotic separation is clear. Neobiogeographers may have similar problems, as the recent discovery of the coelacanth *Latimeria* off the north-east coast of Sulawesi (Erdmann *et al.*, 1998) shows. Because of the lack of fossil representatives for the intervening 80 million years, this fish is one of the most famous so-called 'living fossils'. A new species name, *L. menadoensis*, has been proposed by Pouyaud *et al.* (1999) for this Indonesian specimen on the basis of both morphological characters and DNA sequence divergence. The fish was caught 10,000 km east of the Comores archipelago, where the genus *Latimeria* was previously believed to be endemic with the species *L. chalumnae* Smith. The real specific separation is plausible but needs to be confirmed by further studies on additional specimens, although the possibility that the two populations could be part of a metapopulation seems to be rejected by Holder *et al.* (1999). In any event, the disjunct distribution of the coelacanth raises problems of both the lack of knowledge of its possible distribution in the intervening areas and the biogeographical interpretation (Forey, 1998; Holder *et al.*, 1999).

The latter point involves the definition of the boundaries of a species range, and the patchy distribution recalls the concept of the metapopulation. Distributional expansions or contractions of a fossil species could occur within its geological time range, and insufficient resolution of stratigraphic correlation over a wide geographical area would prevent correct interpretation of the pattern observed.

Disjunction is usually regarded as a historical phenomenon and the causes of disjunct distributions have a particular attraction in biogeography because the different biogeographic models challenge each other. Is the distribution a reflection of chance dispersal events that bring founder populations beyond the pre-existing range of a species ('jump dispersal'), implying the crossing of a selective 'filter', or barrier? Or does it represent vicariant disruption of a pre-existing range?

In neobiogeography and in palaeobiogeography these questions may generate highly speculative narrative hypotheses that must be tested using analyses of both the distributional patterns of other organisms and geological data. Both the taxonomic level considered and the whole history of the biogeographical distribution of the taxon concerned are important for the interpretation of the pattern. The disjunct distribution

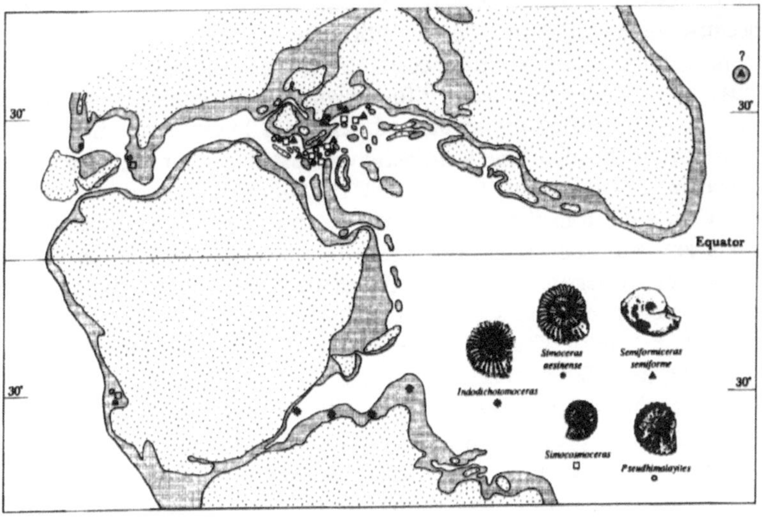

Figure 2.4 Palaeobiogeographical distribution of taxa from the ammonite Semiforme Zone (Early Tithonian). Note the disjunct distribution of *Simocosmoceras*, *Simoceras aesinense* and *Pseudhimalayites steinmanni*.

Source: Cecca (1999). Reproduced with permission of Elsevier Science.

of populations belonging to the same species may be the result of processes different from those responsible for the disjunct distribution of representatives of, say, the same genus or family.

Long time-intervals and processes of continental drift are involved in the disjunct distributions of taxa of high taxonomic rank and typically the vicariant model is invoked in such cases. By contrast, the disjunct distribution of species belonging to groups characterized by rapid evolution, ammonites for example, may require different explanations.

Fig. 2.4 shows the disjunct distribution of conspecific populations in the same relatively short period. Short in this sense is relative to processes acting on a geological time-scale but is certainly extremely long relative to any biological time-scale. The time-interval corresponding to the duration of an ammonite zone is some hundred thousand years. The geographical range of the species *Simoceras aesinense* is considerably larger than that of congeneric species, older and younger, which only occur in the Mediterranean area. This distribution therefore describes a range expansion of the genus *Simoceras*. The disjunct distribution of *S. aesinense* cannot be explained by the vicariant model, because there was no specific separation, unless South American specimens belong to a different species not recognizable by palaeontological means. Either the species is found where it first came into existence as a result of dispersion, as defined below, or longer-term changes in distribution through time were achieved essentially through the same processes of dispersion (Rosen, 1988c). This slow process can obviously take place in the time interval of an ammonite zone. Assumptions are required about dispersion at the larval stage, even about modes of dispersion of organisms having direct development (like cephalopods), because settlement in the intervening areas was seemingly impossible. The distributional pattern of *Simocosmoceras* is different, as different species or subspecies have been described in different localities. A vicariant model can be invoked, but biogeographers might also interpret this pattern as a manifestation of dispersal across a barrier (an unsuitable area) because the ancestors of *Simocosmoceras* are reported only from the Mediterranean. However, dispersal implies the assumption of a 'centre of origin' that is usually impossible to demonstrate. In addition, the term 'centre of origin' reflects different concepts and has been employed in different ways by a variety of authors (see Chapter 3). The case of *Pseudhimalayites steinmanni* (Fig. 2.4) highlights these difficulties because this is a monospecific genus whose ancestor belongs to the pandemic genus *Aspidoceras*. Thus, the range of the family Aspidoceratidae is not modified by the appearance of *Pseudhimalayites* and the processes that explain its disjunct distribution are probably the same as those causing the distribution of *S. aesinense*.

It is worth noting that Norris (2000) has suggested that species' ranges of pelagic taxa are controlled more by limits on their ability to maintain viable populations than by their ability to disperse.

2.1.3.1 Relict distribution

Many disjunct populations are isolated from the areas where their closest allies live because they have been left behind when their ancestral population, once widespread, retreated because of changes in environmental conditions. These populations (or taxa) are called **relicts**. Examples of marine invertebrate relicts are found in European seas, where populations present in isolated localities of the modern Boreal zone are remnants of species that were more widespread during the glacial period, and are now separated from the main populations in the arctic region. Living populations of marine organisms characterized by amphitropical, or antitropical, distributions have also been interpreted as relicts (Briggs, 1995).

2.1.4 Cosmopolitan and pandemic distributions

The term cosmopolitan is widely used and generally indicates wide geographical distribution of taxa that occur in essentially every part of the world where their habitat requirements are met. It indicates the capacity of a taxon to have a worldwide distribution. The term should, therefore, be restricted to taxa whose distribution covers all the habitable areas of the globe relative to their life-environment and obviously a terrestrial organism cannot live in the oceans. The term cosmopolitan is also frequently used in a comparative sense: group A is more cosmopolitan (more widespread) than the group B. The adjective **pandemic** is applied to taxa distributed over very large, but not global, areas. This distributional pattern indicates the possibility of a relatively easy dispersion and also ecological tolerance (eurytopism). The term **semi-cosmopolitan**, also used in the biogeographical literature is considered here as a synonym of pandemic.

2.1.5 Provincialism

This term is also very common in biogeography and is generally used to imply a distribution contrasting with that of cosmopolitan. It means that in certain periods it is possible to identify biogeographical provinces, containing faunal assemblages that are endemic. Provincialism usually implies a restriction of faunal exchanges, relative genetic isolation, and eventually the appearance of new taxa (in a macroevolutionary sense), characterized by a more restricted (provincial) biogeographical distribution than the parent taxa that were more cosmopolitan. It is worth noting that it is always possible to identify provinces, even in the absence of endemic forms (Valentine, 1973). Dommergues and Marchand (1988) differentiated a primary and a secondary provincialism. The former identifies periods without any faunal communication between distinct biotas totally separated by barriers, whereas in the latter faunas are able to merge in areas of ecological overlap.

2.2 The three main categories of biogeographical processes

Classical biogeographical models traditionally emphasize a single process (dispersal, vicariance, ecology, etc.). However, the reality appears to be more complex and different concurrent processes are involved to

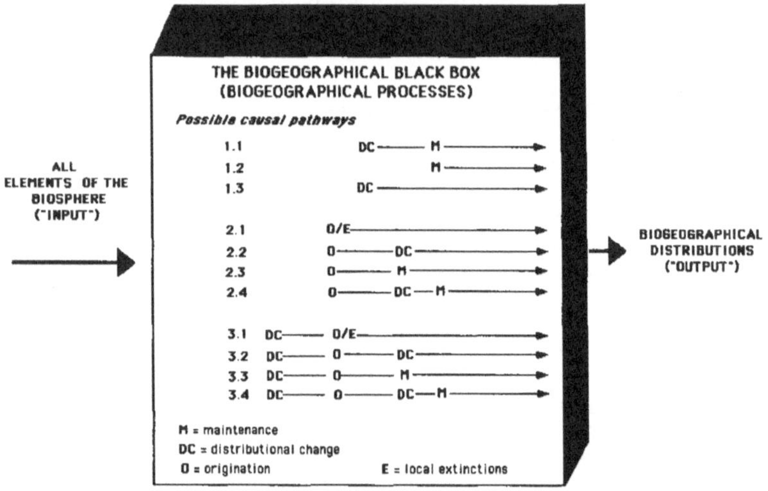

Figure 2.5 Rosen's biogeographical black box. Processes of origination are assumed to be decoupled from geographical and geological factors, or geographical factors are largely fortuitous, in pathways 1.1–1.3 whereas pathways 2.1–2.4 assume that geographical factors and origination processes are interconnected. Note that 'distributional change' includes movements of terranes with their biotas or movements of biotas independently of terranes.

Source: Rosen (1992). Modified with permission of Elsevier Science.

produce biogeographical distributions. These different processes are complementary and not mutually exclusive and some are regarded as distinctly biogeographical processes rather than extended aspects of ecology, evolutionary biology or geology.

Because biogeographical processes are largely conjectural, Rosen (1992) proposed to envisage them collectively as a 'black box', that he called the **biogeographical black box** (metaphorically intended as an 'apparatus of unknown internal design whose internal mechanism is hidden from or mysterious to the users' (Rosen, 1992, p. 173)). This can be regarded as a biological system that generates biogeographical distributions. The processes within the box are those in which the organisms react to each other, and to their physical environment, on a geographical scale. All elements of the biosphere can be considered as the 'input' and are processed within the black box to produce biogeographical distributions that represent the 'output' (Fig. 2.5).

Within the black box, the possible causal pathways for biogeographical distributions are related to the three main kinds of explanations:

1 maintenance processes
2 processes of distributional change
3 origination processes.

Before summarizing Rosen's definitions of these three main categories of processes, it is necessary to note that many of the relevant ideas of processes that biogeographers have imagined to explain distributional patterns can be considered as subsets of the three major categories above. These processes are neither proven nor theoretically necessary to biogeography but are included by him because they represent the biogeographical vocabulary. Numerous ideas of processes have been listed by Rosen (1988a, 1988b) in a sort of 'agenda' that does not imply critical appraisal of their rigour, testability or general acceptance. Some of them have an interface with Earth history and are widely adopted for palaeogeographic reconstructions. These different concepts, processes and ideas of processes believed to be important parts of the biogeographical system will be recalled under the appropriate kind of explanations for biogeographical distributions (maintenance, distributional change and origination).

2.2.1 Maintenance

Under this heading are grouped ecologically-related (though on a geographical scale) processes that actively maintain the geographical extent of the total living populations of a particular species (or conversely, that operate to exclude it), together with those which effectively determine their overall survival within their range areas. They constitute a potential equilibrium between a taxon, its habitat and the other components of its community: in other terms, an equilibrium between autecological and synecological dynamics. For palaeobiogeographical purposes, maintenance processes are not observable because they operate at a short-term time-scale. They therefore operated contemporaneously with the fossil organisms concerned, comparably with palaeoecological processes.

Some of the different ideas included in maintenance processes are listed in Table 2.1. In columns 2 and 3, processes, concepts and ideas are distinguished according to the short or long time-scale at which they operate in biogeography. In columns 4 and 5 the interface with Earth history and the practical value for palaeogeographic reconstructions of some of these ideas is shown. There are ecological processes that are not confined to 'ecological biogeography' and maintenance ideas having an Earth history component which are taken into account under

Table 2.1 Maintenance concepts and processes

	Ecological biogeography: short-term processes	*Historical biogeography: long-term processes*	*Earth history component*	*Practical value for reconstructing Earth history*
1. Dispersion	*			
2. Geoecology	*		*	
3. Equilibrium theory and island biogeography	*			
4. Availability of habitable area	*			
5. Barriers	*		*	
6. Metapopulations	*			
7. Geoecological assemblages	*			
8. Provinces and provinciality	*	*	*	
9. Species-energy hypotheses	*	*	*	

distributional change. Some ideas of processes are discussed here while others (dispersion, barriers) are treated in other parts of this book.

Geoecology consists of (1) the behavioural and ecophysiological processes by which a taxon maintains its geographic range and (2) the environmental factors that favour its presence within its range, with feedback relationships between the two. Ecophysiology acquires a geographic dimension only when related to the geographical limits of a taxon's environmental tolerances. Environmental processes operate from microhabitat to global scale and there is no evident limit between which ecological factors can also be called geographical (Blondel, 1979; Rosen, 1988a); they are relevant in biogeography when they operate at the larger scales. The term 'geoecology' refers therefore to the geographical scale of ecology which is useful to distinguish from the ecological processes themselves when applying fossil distributions to palaeogeographic reconstructions. Geoecological assemblages contain a fossil signature of the biotic response to environmental factors and may be used in palaeogeography.

2.2.2 Distributional change

The most important processes are listed in Table 2.2 according to Rosen (1988a). Some of them are discussed in other sections of this book.

Table 2.2 Distributional change concepts and processes

	Ecological biogeography: short-term processes	*Historical biogeography: long-term processes*	*Earth history component*	*Practical value for reconstructing Earth history*
1. Range expansion	*	*		
2. Jump dispersal	*	*	*	
3. Equilibrium theory and island biogeography	*	*	*	
4. Earth history	*	*	*	*
5. Environmental tracking as a response to:				
5.1. ecological change (including climatic change)	*	*	*	
5.2. the effect of Earth history on ecological change	*	*	*	*
5.3. disturbance	*	*	*	*
5.4. availability of habitable area	*	*	*	
6. Adaptation of a taxon to different conditions	*			
7. Species–energy hypotheses	*	*	*	

Distributional change means the change in the distributional limits of the whole geographical population continuum of the taxon through time, whether the result is expansion, contraction, disjunction or any other kind of regional adjustment. Distributional change is a possible outcome of the response of taxa to perturbations or disruptions of the ecological equilibrium (hurricanes, volcanic eruptions, etc.) constituted by maintenance processes as it may occur at short term. This fuels the long-standing debate about opposite models of equilibrium or disturbance in ecology (see section 7.6.). The physiological constraints of an organism provoke the progressive shift of the margins of distribution if change to environmental conditions, particularly temperature, suitable habitats and food supply, occur. The taxon gradually moves to new areas in order to find environmental conditions compatible with the limits of its physiological tolerance. Range expansions or contractions may be the results of biotic responses to changes of the ecological regime (food supply, competition, environmental parameters, etc.). Disjunction may also occur if the original area undergoes unfavourable conditions: a given species can be fragmented in different populations occupying disjunct areas, although the total distributional area of the species and its enveloping boundaries may remain the same. Rosen (1992) stressed that distributional change is only one of two possible responses to loss of equilibrium. In fact the organism's endogenous processes themselves accommodate, perhaps genetically, a certain amount of environmental change. In this case the distributional area might remain the same. Other possible responses are origination or extinction processes (see below).

Distributional change can occur in different ways, often invoked by dispersalist biogeographers, especially in connection with the ideas of 'centre of origin'. Despite its equilibrium constraints, a given taxon can settle by '**chance**' or '**jump dispersal**' in new suitable areas that were previously separated from the original area by a barrier (e.g. an unfavourable area). The modern cases of distributional change are considered as dispersions as well as chance dispersal (anemochorian, zoochorian transport, rafting, etc.).

Some processes of distributional change can be directly observed in neobiogeography because they can also operate on the same time-scale as maintenance processes (e.g. changes in response to climate changes). They correspond therefore to cases of **dispersion**, as defined below.

Distributional change may also occur on much longer time-scales (undetectable on human time-scale), as shown by responses to plate-tectonics (see section 2.2.2.3.). In palaeobiogeography processes of distributional change can only be inferred and must be intended in a pure historical sense: they may have happened and then ceased in the distant past.

2.2.2.1 Dispersal and dispersion: a useful distinction

The former term is the most popular of the two in biogeography and commonly indicates different biogeographic processes acting at different temporal scales such as the larval transport over great distances, the crossing of barriers, or the expansion of the distributional area of a group with processes of species origination. The very broad use and the different meanings of the term 'dispersal', particularly the lack of distinctions between long-term and short-term acting processes, highlight the confusion existing in most dispersalist explanations of distributions.

Rosen (1988c) thoroughly discussed the different meanings of 'dispersal' and favours the term **dispersion** (in analogy with the physical behaviour of gases) in discussing the latitudinal distribution of reef corals. In particular, dispersion implies the propagation of an organism over a certain geographical distance, which is a function of both vagility and larval biology (see Chapter 6), and its final recruitment. Different modes of dispersion may occur as demonstrated by modern, observed, cases of 'chance dispersal'. Dispersion is a process acting on a 'biological' (short-term) and not 'geological' temporal scale; it maintains the genetic exchange between populations of widespread species even over great distances (Scheltema, 1979; Jablonski and Lutz, 1983). A reversed distinction of dispersion and dispersal was also proposed by Armstrong (1977), i.e. the use of 'dispersal' for short-term and 'dispersion' for long-term processes.

Lieberman and Eldredge (1996) introduced a different type of dispersal called **geodispersal**, whose definition was later extended by Lieberman (2000, p. 79) as 'the expansion of the range of a group of species due to the elimination of some topographic or climatic barrier . . . followed by the emergence of a new barrier which produces subsequent vicariance'. A temporal succession of events is therefore involved:

1 a geographical barrier, either geological or climatic, separates two distinct biotas
2 later the barrier is temporarily eliminated (for example by sea-level rises or plates collisions), allowing formerly separated populations of different clades to merge and to expand their ranges simultaneously into new areas
3 subsequently a barrier rises again to produce allopatric speciation and vicariance in several taxa.

Geodispersal does not appear to introduce new concepts in biogeography because the second event corresponds to a pattern already described by Hallam (1974) as biogeographic convergence (see section 2.2.2.3) and the third event corresponds to vicariance.

In this book the term 'dispersal' is understood in an entirely historical sense and it is synonymous, on a long-term time-scale, with distributional change. The latter term is preferred and will be used in this book according to Rosen (1988c, p. 285). It is worth noting that dispersal or distributional change, as accepted here, is not an entirely 'active' process because areas moved, and are still moving (e.g. through tectonism), therefore the biotas do not move only with respect to a static geography. Plate movements can bring areas within the dispersion range of particular species, thus producing distributional changes caused by convergence instead of jump dispersal. The same applies to Earth history events, precisely tectonic, eustatic, climatic and oceanographical events, abbreviated as TECO events by Rosen (1984). Rosen (personal communication, December 2000) now adds geomorphological, erosional and depositionary events to TECO.

2.2.2.2 Range expansion and the use of the term 'migration' in palaeobiogeography

The movement of a living organism is realized by two short-term processes: dispersion and migration. In particular the term **migration** means that the organisms move by rhythmic and directional processes. The migration can be diurnal, seasonal or may last an entire life cycle and even occur as a 'one-way' migration from the original area with the return realized by the offspring. These kinds of movements cannot be recognized in palaeontology because they are the result of very short-term processes. Nevertheless, migration (a term corresponding to a sum of different processes) is frequently used in palaeobiogeography to describe patterns of change in the geographic distribution of a taxon during long-term time intervals, corresponding to cases of either **range expansion** or **range contraction** associated with shift of the distributional area. Patterns of such a kind are the result of a summation of distinct biotic and abiotic processes acting over both short and long time intervals, and should be included in the category of the distributional changes. Migration, as intended by biologists, is not synonymous with distributional change.

In palaeobiogeography, because of the different temporal dimension, so-called 'migrations' are actually patterns which describe historical outcomes and do not represent the effects of processes acting during the life cycle. Therefore in palaeobiogeography the main problem is the fate of the recruitment of a species outside its 'normal' distributional area and not the reasons of its movement, whatever they could be.

Box 2.1 discusses some of the well known, and thoroughly studied, repeated 'migrations' of Tethyan and Boreal ammonites (but similar patterns can be observed in other organisms) during the Jurassic and the Early Cretaceous.

The palaeobiogeographical patterns of distributional change discussed in Box 2.1 show different evolutionary issues labelled with a disparate nomenclature; they do not correspond to real processes of migration. This is the reason why I propose to dismiss the use of the term 'migration' in palaeobiogeography in favour of other terms which can better describe the observed distributional patterns of fossil taxa. But what terms? From the descriptive point of view, a part of these patterns could be described as cases of range expansion. The term 'range expansion' indicates that a taxon has colonized favourable areas beyond its initial range, through larval dispersion by marine currents or by normal means of locomotion. This pattern is interpreted by dispersalist biogeographers as the expansion (from a centre of origin) of a taxon's range over a barrier with eventual diversification. This would imply the underlying, erroneous, assumption of a stationary Earth where organisms move by springing from evolutionary centres of origin to colonize new areas. However, it is impossible to ignore the role of plate movements, oceanographic and climatic changes in controlling the distributional ranges of the organisms. Instead, the term 'range expansion' can be used with a purely descriptive meaning to indicate the colonization of a new area driven by processes of distributional change. Crame (1993, 1996) uses the terms 'range expansion' and 'range contraction' to indicate the biogeographical response to repeated climatic changes. The shift of the distributional area is another pattern which can be associated or not with range expansion or contraction. Range contraction with concomitant shift is a pattern which is often associated to the concept of refuge (Taviani, 1998).

Range expansion/range contraction and migration (as intended in palaeobiogeography) describe patterns and do not identify processes, which correspond to environmental tracking (see below). We need to distinguish on one hand the biotic movement on geological time-scale towards regions out of the common distributional area, and on the other hand the evolutionary issues. I propose to include within the category of 'distributional change' the following concepts.

Box 2.1 Are 'migrations' strays, invasions, mass occurrences or spreads?

Termier and Termier (1952) used the term 'transgression' to indicate range expansions of faunas. In discussing the southward range expansion of the Boreal ammonite family Cardioceratidae, Arkell (1956, pp. 610–611) used the term 'boreal spread' instead of 'transgression' to keep distinct the geological concept of marine transgression from that of faunal 'migration'. I understand Arkell's use of the latter term as a pattern of distributional change implying geological (long-term) time intervals.

Tintant (1972, in 'Congrès sur les méthodes de la stratigraphie', p. 43) correctly distinguished the real biological process of migration from what he called '**transgression animale**'. He regarded the term 'migration' as corresponding to the change of the distributional area of a species due to short-term ecological variations of its life-environment. On the other hand Tintant's 'transgression animale' corresponds to a process of 'range expansion', i.e. an increase of the distributional area of successive species of a genus or a phyletic line. The example proposed by Tintant is the progressive southward expansion of different successive species of the Boreal ammonite genus *Kosmoceras* during the Callovian. Changes of the distributional area of successive species of a genus may also result in a complete shift of the species range that does not entail range expansion.

Enay (1980b) and Enay and Mangold (1982) used the term 'migration' to describe patterns which are the results of processes that occurred over long-term time intervals. The authors analysed the major episodes of distributional change of Jurassic ammonites, stressing the different evolutionary fate of expanding Tethyan and Boreal taxa. Colonization, i.e. evolution of new lineages in the new areas (see section 2.1.1.1), characterizes Tethyan spreads ('migrations' in the French text of Enay and Mangold), whereas Boreal ammonites did not colonize the southern areas, the main evolutionary events occurring in the Boreal areas. For the latter case Enay and Mangold used the term 'spread' ('expansion' in the French text) as opposed to migration. It is worth noting that opposite evolutionary results characterize episodes of distributional change of Tethyan and Boreal ammonites.

The changing relationships between Early Cretaceous ammonite faunas of Tethyan and Boreal origins in Western Europe led Rawson (1973) to distinguish **isolated strays** from **mass migrations**. The case of isolated strays is illustrated by the example of Tethyan Valanginian

Box 2.1 continued

and Hauterivian ammonites (Rawson, 1993) which came into the Boreal regions singly by chance migration or penetration in an unfavourable environment, or even by post-mortem drift of empty shells. These ammonites occur randomly through the stratigraphic succession as 'short-lived' (but I argue that it is for an interval which exceeds in any case the biological time-scale) populations that did not evolve into new species and remained for a brief period, appearing at defined horizons often associated with other 'immigrants'. The 'afflux téthysiens' (Cecca, 1988) or 'Tethyan spreads' (Cecca, 1999) are sporadical range expansions of Tithonian ammonites which had their maximum development in distal pelagic habitats of the Mediterranean Tethys, on to the European shelves under the control of sea-level fluctuations and different ecological factors. These spreads correspond to Rawson's 'isolated strays' and probably resulted in short episodes of recruitment in new areas, without eventual colonization. They are also equivalent to Jurassic spreads of Boreal ammonites. The Late Oxfordian 'short-term invasions' of the Boreal ammonite genus *Amoeboceras* (family Cardioceratidae) in Central and Western Europe, described by Matyja and Wierzbowski (2000), is another example of an isolated stray. Compared to conspecific forms of the Boreal regions, the 'invasive' forms show smaller shell-sizes and crowded septa in a long interval of the last part of the phragmocone. This has been interpreted as a biological response leading to the attainment of a long period of sexual maturity due to different environmental conditions, but *Amoeboceras* did not colonize the Submediterranean areas because it did not give rise to new species there.

Mass migrations, renamed **regional mass occurrences** by Kemper *et al.* (1981), correspond to faunas which arrived in considerable numbers from Tethys to form evolving populations that normally temporarily displaced Boreal faunas. These taxa can either occupy newly formed niches or compete successfully against pre-existing forms and thus evolve *in situ* to form endemic species or supraspecific taxa. Regional mass occurrences are equivalent to Tethyan spreads of Jurassic ammonites and resulted in colonization of the new areas. Kemper *et al.* (1981) also recognize **inter-regional mass occurrences** when a Tethyan fairly long-ranged genus suddenly expanded dramatically in numbers in both north-west Europe and Tethys. These authors cited genera *Olcostephanus* and *Bochianites*, whose occurrence in Boreal areas has been interpreted by Cecca (1998) as opportunistic behaviour of generalist taxa.

1 **Faunal spread** indicates the long-term biotic movement towards new areas; it is inferred from patterns commonly indicated with the terms 'dispersal' and so-called 'migration'; **faunal retreat** corresponds to the opposite process inferred from range contraction patterns.
2 **Unsuccessful recruitment** identifies an evolutionary issue. It indicates that no colonization has taken place after the change of the distributional range; it is inferred from patterns indicated with the terms 'isolated stray', 'invasion' (see Box 2.1). The case of 'pseudopopulations' (see section 2.1.1.1) is an example of unsuccessful recruitment on a short-term time-scale.
3 **Colonization** (or **successful recruitment**) indicates that evolution of new lineages, specific and supraspecific, has taken place in the newly colonized areas; it is inferred from patterns indicated with the terms 'range expansion', 'mass occurrence', 'mass migration' and partly 'dispersal' and so-called 'migration'.

The identification of genuine migrations in the geological record is absolutely speculative: it can be suspected, at best, in the case of disjunct distributions of conspecific taxa. The distribution of the ammonite species *Simoceras aesinense* (if the specimens recorded in the Mediterranean areas and in Cuba, Mexico and Argentina are really conspecific) discussed above (Fig. 2.4) could be interpreted as a trace of real migrations whose historical succession and periodicity (if any) is obviously not detectable from the geological record. The disjunct distribution of the same species could represent either a cumulative (relative to time) record of migrations or dispersions with successful or unsuccessful recruitment.

2.2.2.3 Geology and processes of distributional change

During the last quarter of the twentieth century all previous views of global geological processes were profoundly modified by plate tectonics theory. This also had a dramatic impact on biogeography, giving scope for application to recognition of displaced terranes and also arguments to explain the distributional patterns of living and fossil taxa, particularly in the light of the vicariance model.

The relation between distributional change and geological processes is crucial in palaeobiogeography. The changes of the continental positions and eventual suturing and rifting events result in distributional changes whose major outcome is the change in the taxonomic composition of the biotas at different places through time. Hallam (1974) illustrated the changing patterns describing two major historical phenomena: **convergence** and **divergence**.

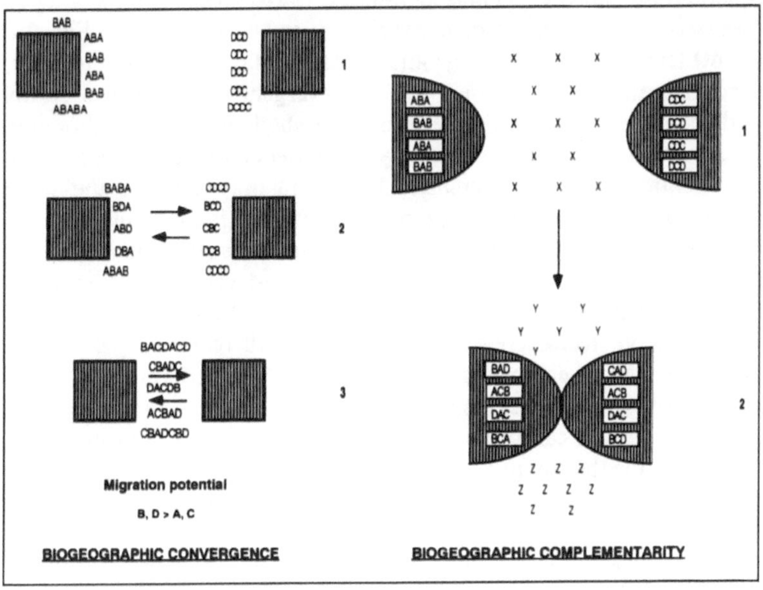

Figure 2.6 Theoretical schemes illustrating the concepts of biogeographic convergence and complementarity. Continents are represented by vertical lines. In the left scheme, as the continents approach each other as a result of plate movements the faunas (A, B, C, D) progressively merge. In the right scheme terrestrial (A, B, C, D) and marine (x, y, z) faunas show an opposite behaviour as continents approach each other and eventually become sutured together. The terrestrial faunas merge and exhibit convergence because a corridor has been created, which corresponds to a barrier for marine faunas which diverge from the common ancestor x to vicariant faunas y and z. Both schemes can be read in the opposite direction.

Source: Hallam (1974). Modified with permission of Blackwell Science.

Convergence indicates intermixing of biotas due to continental shifting: as two continents approach each other their associated marine shelf faunas progressively merge, with a time sequence depending on migration potential (Fig. 2.6). **Divergence** corresponds to the converse phenomenon which occurs when continents move apart (or when two land areas are isolated by a seaway). The growing increase of provincialism related to the opening of the Atlantic Ocean is a classic example of divergence (see Fig. 4.5).

Convergence does not correspond to the evolutionary concept of morphological convergence of phyletically unrelated taxa, and therefore it should be indicated as **biogeographical convergence**. The merging of biotas is the final effect of convergence: in the example of Fig. 2.6 (right) terrestrial biotas merge and marine biotas diverge. Conversely, a continental separation may lead to divergence of terrestrial biotas and convergence of the marine ones. Very shallow water and shoreline biotas are also affected by deep water barriers and will diverge as the continental separation increases. Eustatism may produce the same effects as plate tectonics: for example, a sea-level fall may produce the disappearance of a shallow seaway allowing terrestrial biotas to merge and marine biotas to diverge, whereas a sea-level rise may create a seaway and provoke opposite biogeographical effects.

A single geotectonic event can produce both biogeographical convergence and divergence. Hallam (1974) designated this pattern as **complementarity**. For example, land connections and convergence of terrestrial faunas can occur as a result of continental suturing but at the same time the communication of marine faunas can be interrupted, enhancing divergence (Fig. 2.6). Classic examples are the establishment of the Isthmus of Panama, which created in the Pliocene a land corridor between North and South America, and the collision of Africa and Arabia with Eurasia in the Miocene. In the first example two formerly distinct terrestrial biotas progressively merged as the land communication was established (Simpson, 1950; Webb, 1991), while marine faunas became isolated in the Pacific and Atlantic areas giving rise to distinct species. The latter event can be described with the vicariant model. The second example corresponds to the isolation, and subsequent divergence, of marine faunas that previously were part of a single Tethyan biota and the correlative biogeographic convergence of terrestrial biotas, with elephants, hippopotamuses and rhinoceroses spreading from Africa into Asia and Europe (Steininger and Rögl, 1984; Briggs, 1995).

The geological processes might produce distributional change in different ways. McKenna (1973) has distinguished two basic processes that he respectively called 'Noah's Ark' and 'Beached Viking Funeral Ship'. The first one corresponds to contemporaneous transport of living biotas on a moving terrane; the second one corresponds to the transport of fossil biotas. Australia gives good examples of both processes as it is now docking against South-East Asia. Other examples may be found in the palaeogeographical history of the numerous terranes of Gondwanan origin which were accreted to Asia during the Mesozoic and Cenozoic (Dercourt *et al.*, 1993).

2.2.2.4 Island integration

On the basis of reconstructions demonstrating the integration into the Hawaiian-Emperor chain of some islands and seamounts which originated outside the present Hawaiian hotspot (e.g. Wentworth Seamount or the submerged northern portion of Necker Island), Rotondo *et al.* (1981) proposed the mixing of two different biotas, Hawaiian and non-Hawaiian, as a mechanism in the formation of Hawaiian Island endemism. It is worth noting that during the last 38 million years between 50 and 60 pre-existing older seamounts were incorporated to the Hawaiian chain. **Island integration** is therefore a possible alternative model to founder principle, particularly for terrestrial and littoral components of the biota. According to Rotondo *et al.* (1981), the high degree of primary endemism recorded in some islands could not be explained by the unusual dispersal events required by the founder principle (anemochory, hydrochory, biochory, etc.). Island groups formed on the Pacific plate became deeply submerged or remained emergent as volcanic islands, thus becoming isolated. The isolation led to divergence and then to speciation. Non-Hawaiian islands shared a biota that was not present at/on the emergent Hawaiian Islands of the time. This model, which is conceptually similar to 'Noah's Ark' except for the scale of the geological elements concerned, is supported by the affinities with faunal elements of other islands on the Pacific plate.

Rosen (1992) has discussed other possibilities, as follows.

- A taxon may occupy a plate (or a terrane) that remains geographically static with respect to the Earth but it may be subjected to climatic change. The taxon's response may result in distributional change.
- The terrane occupied by a taxon moves with respect to latitude and/or longitude, or also vertically by subsidence (isostasic processes, rifting, etc.), and brings the taxon into different external conditions (climate, depth, altitude, etc.): for example, oceanic volcanic islands which slowly become submerged as the plate on which they ride subsides.
- The terrane moves but the taxon also moves to keep the same distributional area with respect to a global reference point (e.g. a pole of rotation), thus they move relatively to each other.

It is worth noting two aspects.

1 These cases of distributional change refer to long-term time-scales and therefore the biogeographical analyses apply to taxa of high taxonomic ranks.
2 The terrane movements mainly (but not only) affect terrestrial organisms because the marine ones have higher potential of dispersion; even the dispersion of benthic faunas characterized by non-planktotrophic larval stages is quicker than terrane movements. In other terms, the biological response to the creation of, say, a seaway is quicker than the geological process itself.

2.2.2.5 Environmental tracking

Consider the distribution of a taxon in a particular area at a given time determined by ecological barriers. If the geographical extent of the ecological conditions change, the distribution of the taxon passively changes in time. This is the **environmental tracking as a response to ecological change** (Fig. 2.7). The species either does not change its distribution at all or becomes extinct, therefore at least a part of the new suitable area must overlap geographically with the previous one long enough for the species to 'follow' the environmental shift (Rosen, 1988a). An example is provided by distributional changes of faunas during the glacial and interglacial phases of the Quaternary.

Depending on time-scale and cyclic character, it is impossible to discriminate the palaeobiogeographical effects of changes in ecological conditions. However, long-lasting changes or particular historical events may result in biogeographical changes of geoecological assemblages or bioprovinces through time (faunal spread, faunal retreat). The different effects of environmental tracking are identified by patterns such as range expansion (here with implications for the centre of origin ideas), contraction (the idea of refugia) and divergence or convergence. Successful or unsuccessful recruitments are the possible evolutionary issues of these cases of distributional change (so-called 'migrations'). The eustatically controlled arrivals of Tethyan ammonites, and other components of the epioceanic biota, on to the peri-Tethyan shelves during the Mesozoic can be described as cases of environmental tracking due to ecological change. These expansions of the ecospace of organisms adapted to live off the shelf-break brought them to enter areas of former epicontinental platforms where environmental conditions (currents, temperature, nutrients, etc.) driven by sea-level rise modified physical and ecological parameters (Oloriz *et al.*, 1991, 1993). Successful recruitments are documented in the Jurassic and have been interpreted by Oloriz (1986, 1990) as the result of the

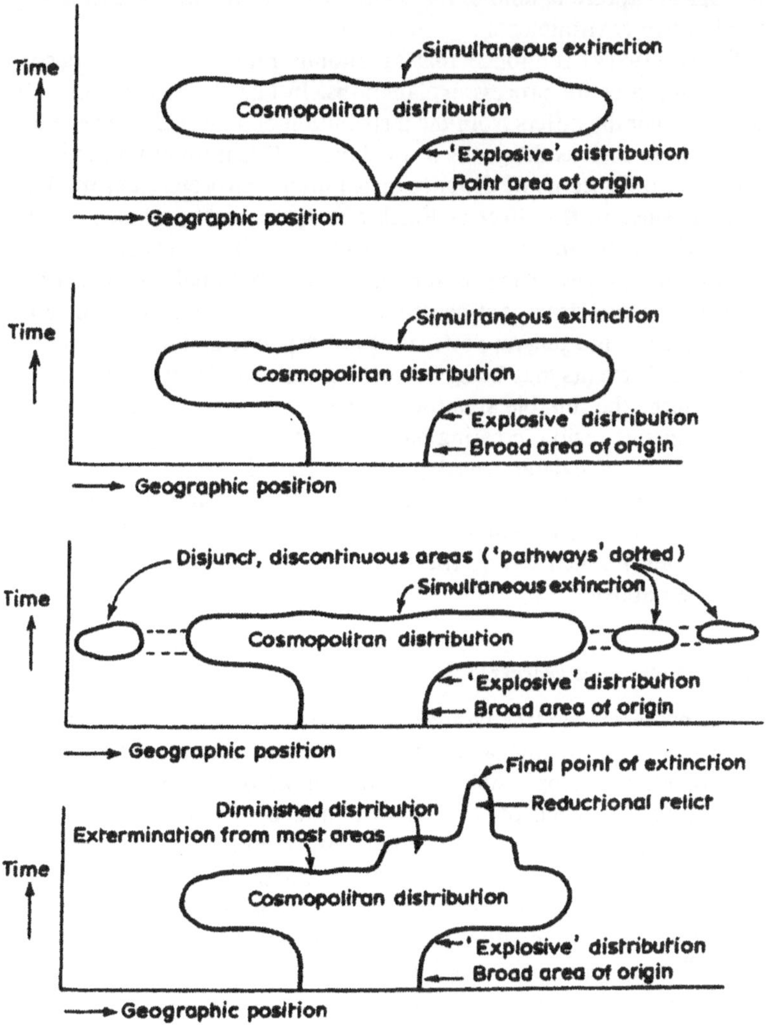

Figure 2.7 Environmental tracking as a response to ecological change. The different patterns of distributional change through time may correspond to the passive change of the range of a taxon because of changes in the geographical extent of the ecological conditions which favour it. Note that a dispersalist explanation is not necessary for the patterns illustrated: range expansions reaching their acmes with the phases of cosmopolitan distribution can be passive responses to ecological changes and not actively determined by dispersal.

Source: Boucot (1975b). Reproduced by permission of the author.

ecological capture of ammonites which gave rise to new forms adapted to platform environments.

Rosen (1988a) cautioned that taxonomic rank may imply conceptual differences for process explanations. In fact the distribution of a species generally differs from the distribution of a more inclusive, high-rank, taxon such as the genus or the family. This is important in interpretation of patterns of disjunct distribution. **Environmental tracking as a response to the effect of Earth history on ecological change** corresponds to Earth history events (Rosen's TECO events) driving ecological changes which in turn drive distributional changes. Plate movements may drive continents into a different climatic belt, or cause the change of the general patterns of atmospheric and oceanic circulation. TECO events may bring two ecologically similar regions together and enlarge the suitable area for organisms, or subdivide a previously homogeneous region, creating barriers to biotic exchange.

The term of **habitat tracking**, particularly invoked in the framework of the 'punctuated equilibria' theory as a possible biogeographic response to environmental change (see Eldredge, 1995), is a more general term which covers the two types of environmental tracking discussed above.

2.2.3 Origination processes

These processes can be envisaged as possible responses to perturbation or disruption of maintenance equilibrium. The creation of new species having different ecophysiological constraints from the parent species is consistent with concepts of allopatric and peripatric speciation that clearly consider a geographic component. However, if the disruption of environmental equilibrium produces excessively harsh life conditions for a given species, a possible outcome is extinction.

The timing of origination processes is variable, as also discussed for the different types of speciation (see Chapter 1), but it can be considered slow relative to maintenance processes. As regards palaeobiogeography, there is no doubt that origination processes should be regarded as 'historical processes'.

The ideas of origination processes are listed in Table 2.3. Origination is not entirely synonymous with speciation because some envisaged processes of speciation do not require a geographical dimension, although they appear to be rare in comparison with geographic speciations. Sympatric models of speciation imply that origination is not a biogeographical process at all. Origination may also be responsible for macroevolutionary patterns, i.e. creation of supra-specific taxonomic

Table 2.3 Origination concepts and processes

	Ecological biogeography: short-term processes	*Historical biogeography: long-term processes*	*Earth history component*	*Practical value for reconstructing Earth history*
1. Empirical phylogenetic biogeography		*	*	*
2. Vicariance as a response to:				
2.1. Earth history (including climatic change)		*	*	*
2.2. ecological change or the effect of Earth history on ecological change		*	*	*
3. Fringe isolation as a response to:				
3.1. ecological change (including climatic change)		*	*	
3.2. the effect of Earth history on ecological change		*	*	
4. Jump dispersal		*	*	
5. Depth parapatry		*	*	
6. Seasonal sympatry		*	*	
7. Centres of origin (in part)		*	*	
8. Geoecology through time		*	*	*

categories, particularly families, orders, etc. In this case we refer to patterns and not to processes that primarily apply to species (Rosen, 1992).

In allopatric and peripatric models of speciation, a causal geographic factor is needed for division and isolation of an ancestral population and its gene pool. Different mechanisms have been proposed by authors; most of them are discussed and developed in other parts of this book:

- 'dispersal', implying dispersion followed by settlement of founder populations and then colonization of new territories beyond a barrier
- vicariance through either break-up and then drifting of terranes transporting biotas ('Noah's Ark') which are isolated from a common ancestral biota or division of an ancient area by creation of a mountain chain etc.; vicariance may be therefore a response to TECO events
- peripheral, fringe, isolation of fractions of an ancestral population in areas with different environmental conditions with subsequent speciation if geographical isolation of the subpopulations (and their gene pools) persists; ecological change may be also driven by geological factors.

The underlying processes of these mechanisms of genetic division are processes of distributional change. According to Rosen (1992), it is not necessary to invoke either terrane movement or large-scale geographical change to bring about isolation: this process may be provoked, for example, by changes in oceanic currents (temperature, nutrients, etc.) or climatic change. As regards pelagic organisms, palaeoceanographic and climatic change seem to influence rates of turnover by modifying surface water masses and environmental gradients between them to create new habitats rather than by preventing dispersal (Norris, 2000).

Origination processes as well as speciation processes may have implications for the concept of 'centre of origin'. Because a taxon or a group of taxa originated in a particular place, the area of oldest record is particularly important in biogeography as a whole. In palaeobiogeography this is considered the 'centre of origin', although it is objectively impossible to find a causal connection between this pattern (the oldest record of a given taxon in a given area) and the true underlying processes that are responsible of the creation of a taxon. As discussed in another section of this book, the different patterns that are customarily used to infer the location of 'centres of origin' are not independent from the limits of both the fossil record and the taphonomic bias. Therefore the true area of origin of a given taxon cannot be objectively identified. Nevertheless, the recognition of areas of oldest record is useful to describe historical patterns of distributional change.

3 Classical theoretical models: dispersal and vicariance

As far as historical biogeography is concerned, two schools of thought exist about the interpretation of the biogeographical distributions; they are commonly known as '**dispersal biogeography**' and '**vicariance biogeography**'. In the literature dispersal and vicariance have often been used as explanatory models for both disjunct distributions and evolutionary cladogenetic patterns.

If, according to Croizat (1982), Darwin was the first to report distinctly distributions of vicariant types, Croizat himself clarified that he was not the father of the 'vicariance biogeography', a theory which combines Hennig's phylogenetic systematics, Croizat's panbiogeographic method and Karl Popper's scientific philosophy (Nelson and Platnick, 1981). Croizat, whose work did not depend at all on cladistic analysis, constantly expressed strong criticism of Hennig's thought.

Dispersal and vicariance have often been regarded as two distinct biogeographic processes (which in my opinion are actually patterns resulting from both short-term and long-term processes) whose respective role and importance in determining biogeographic distributional patterns provoked a spirited debate between biogeographers of the two schools. It is sufficient to quote McCoy and Heck's (1983, p. 18) reply to Briggs (1981) to have an idea of the rudeness of the debate: 'Current historical biogeographical literature is replete with confusing jargon, personal attacks, and self-serving statements'. This debate began essentially in the 1970s and when the increasing evidence of continental drift brought some biogeographers (Nelson, 1973) to rediscover the works of Leon Croizat (1952, 1958, 1962–1964) on one hand and to develop the concept of vicariance biogeography, based on the use of cladistic analysis, on the other.

I cannot fail to note that the controversy between dispersal and vicariance biogeographers has taken place among neobiogeographers, particularly specialists of terrestrial and/or freshwater organisms, who tried to explain the historical processes that produced the distributions of living organisms. Although palaeontologists, and palaeobiogeographers in particular, have different methodological constraints due to the nature of the fossil record, in this chapter I have decided to expound dispersal and vicariance theories on the basis of the most important neobiogeographers' works. Therefore the reader will not be surprised by my reminders, in brackets, of the definition of temporal scale in the use of the term 'dispersal'.

The controversial differences between schools of thought are not limited to the question of what process is the most important in explaining distributions, but also concern the philosophy, particularly for vicariance biogeographers, of how interpretation should be approached. Vicariance biogeographers hold the idea that dispersal biogeography proposes untestable *ad hoc* explanations, also termed (pejoratively) narrative or 'just so stories', that cannot be falsified. Many vicariance biogeographers claim to adhere to Popperian philosophy and the hypothetico-deductive method and hence the use of the cladistic analysis because it would generate testable hypotheses. In general they state that the first attempt to explain a distributional pattern should consider the vicariance hypothesis because this can be falsified and then may lead, heuristically, to other hypotheses that can approach the reality. These may include dispersal.

The most confrontational of vicariance biogeographers dismiss narrative explanations because they correspond to an inductive approach which has nothing to offer scientifically. Rosen (1988c, p. 270) pointed out that induction is also indispensable 'and in contrast with . . . views of systematists and biogeographers who have adopted a "Popperian" standpoint, it should not be regarded as an unrigorous, uncritical amassing of facts, but a basic part of intelligent pattern-seeking and modelling'. This point of view is supported by Hallam (1994) and seems to be the more realistic, at least in the particular context of the palaeobiogeography we are dealing with.

3.1 Barriers and filters

In biogeography a **barrier** ideally represents a given factor which interrupts genetic flow between populations of an area and isolates them in geographically more restricted areas than the previous one. If the barrier persists for a sufficiently long time-span, the populations can develop into new species. This model clearly derives from the concept of allopatric speciation.

The concept of a barrier is used by both dispersal and vicariance biogeographers, but the two groups give barriers different importance and different roles. The dispersal hypothesis implies that barriers exist beforehand and can be crossed, more or less successfully, by some taxa with possible diversification resulting from isolation in the areas located beyond the barriers. Vicariance biogeographers consider barriers as the means of separating populations that will vicariate by subsequent allopatric speciation. Speciation is therefore contingent on emergence of the barriers.

Barriers can be easily represented with geographic elements that belong to the common observation: mountain chains, rivers, seas, climatic belts, etc. Accordingly, the biogeography of terrestrial organisms is easier for explaining its models through the existence of intervening barriers between different populations and species. Nevertheless, climatic change has been inferred to be involved in the formation of bipolar distribution patterns of marine organisms at different geological periods, which implies an important role of climatic barriers in vicariant patterns (Crame, 1992, 1993).

The most obvious geographical barrier to envisage in the marine environment is the emergence of a land barrier (e.g. the isthmus of Panama at the end of the Pliocene). But marine populations, especially of planktonic organisms, can be isolated from each other by much less tangible barriers. These can be, for example, climatic or represented by abrupt changes in the physico-chemical properties of the water masses, as demonstrated by present-day observations (Norris, 2000). But their recognition in the geological past is generally a matter of speculation, unless they are corroborated by isotopic and palaeontological (key geoecological assemblages) data. In this sense barriers are hypotheses because they are interpretations of distributional limits in terms of particular processes such as ecophysiological ones. In their immediate short-term effect on organisms, barriers are ecological and relate to ideas about maintenance processes. However, they might also be related to major geological features (mountain chains, an isthmus, etc.), i.e. elements having a long-term effect on biotas. Also, depth/temperature limits of stratified water masses may act as barriers leading to speciation, as recently shown in planktonic foraminifers *Globoconella* (Schneider and Kennett, 1999) and *Globorotalia truncatulinoides* (De Vargas *et al.*, 2001) (Fig. 3.1).

In any case, it is demonstrated that some barriers can be effective for some organisms but not for others that, therefore, can cross them. This leads to the concept of **filter**, or **bridge**, to indicate the possibility, for some taxa, to cross the barrier. The first term is mostly used in marine

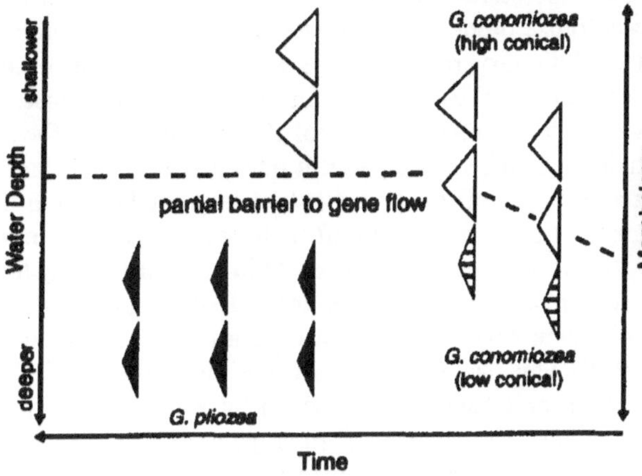

Figure 3.1 Parapatric and allopatric models of depth segregation and speciation within the planktonic foraminiferan genus *Globoconella* of late Miocene–Pliocene age. In the allopatric model of speciation (left) the depth is the reproductive barrier between different morphotypes of *G. (G.) conomiozea* which leads to the origination of *G. (G.) pliozea*. In the parapatric model (right) reproductive isolation occurs between high and low conical morphotypes at opposite ends of a depth-associated morphocline.

Source: Schneider and Kennett (1999). Reproduced by permission of the Palaeontological Society.

biogeography and the second in biogeography of terrestrial organisms (e.g. a strip of land bridging two continents separated by the sea).

3.2 The dispersal model and the 'centre of origin'

The term 'dispersal' has been used in different ways, particularly to indicate processes acting at different time-scales. We do not need now to review the different meanings given to this term by different authors, but concerning the opposition between 'dispersal' and 'vicariance' biogeographies, the literature shows that the term 'dispersal' has been used in a historical sense and also to indicate short-term processes related to propagation and recruitment (transport by wind of winged and plumed seeds, larval transport, etc.).

The central idea of 'dispersal biogeography' is the existence of a particular area where clades of numerous kinds of organisms are supposed to have continually originated, the so-called **centre of origin**. If the clades are successful they spread out from that area, through dispersal, to colonize other environments. The area acts, therefore, as a centre of evolutionary radiation and supplies species to other areas that are less active or less effective in an evolutionary sense. Bremer (1992) uses the term **ancestral area** for the distribution of the ancestor of an individual clade, stressing that it does not need to be limited to a single place. Bremer affirms that if extant taxa are limited in distribution, it does not seem plausible that their ancestor was cosmopolitan. However, the appearance of a new clade in a 'centre of origin' implies an expansion phase relative to its area of origin (see Fig. 2.7), but there is no reason why it should not have an initial contraction phase (Rosen, 1984). This has been demonstrated by Parker *et al.* (1999) for Neogene planktonic foraminifers, which have tended to originate by restricting rather than increasing their geographic range. Therefore Bremer's (1992) statement about widespread ancestors does not have a general value.

According to Nelson (1983), the idea of the centre of origin dates back to Linnaeus; clearly, implications of the more or less unconscious acceptance of the biblical version of creation give scope to historians or sociologists of science. Briggs (1981) stressed that Darwin was the first to write about centres of origin, which he called 'single centre of creation', arguing that each species was first produced within a single region and that it subsequently migrated from that area as far as its powers of migration and subsistence under past and present conditions permitted. Patterson (1983, p. 14) also considers this concept as an 'inescapable corollary of evolutionary (Darwinian) theory'. Adams (1902) provided criteria for the determination of what he called 'centers of dispersal', the most important of which was the location of 'the greatest differentiation of a type'. Matthew (1915) stated that the more advanced species would be found at the centre of origin with the most conservative or primitive species the farthest from it. Darlington (1959) emphasized that centres of origin were places where competitively dominant animals were produced and appeared to be inhabited by the most advanced members of a phyletic group, thus supporting Matthew's theory. Hennig (1966), the pioneer of the cladistic systematics (which is the basis of the cladistic approach to biogeography) stated that species possessing the most primitive characters are found within the earliest occupied part of the area, i.e. the centre of origin.

In summary, it seems that contradictory ideas of 'centre of origin' and different criteria have been commonly considered to determine the centre of origin. The most used are:

1 the area supposed to represent the centre of origin contains the oldest known fossils of the group
2 the area is characterized by the predominance of geologically young taxa
3 the area bears fossils that possess characters regarded as being primitive
4 the area displays the greatest diversity of living species
5 the area is characterized by the presence of those living species that appear to possess the most primitive characters
6 the area is characterized by those living species that appear to possess the most advanced characters.

Criterion 1 is simply a pattern which can merely correspond to the place where the oldest known fossils are preserved: these are not necessarily the fossils of the first representatives of the taxon. On the other hand, further fossil findings in other areas can disrupt the previous pattern. Criterion 4 is also a pattern, in the sense that one can readily define areas of high diversity whose existence says nothing about the processes which might have caused it. Criteria 2 and 3 on one hand, and 5 and 6 on the other, are clearly contradictory.

Amongst dispersal biogeographers there are two different groups:

1 the group that follows Hennig's principles holds the idea that the most highly evolved species are found at the margins of the dispersal area because there they have adapted to new and probably different habitats (criterion 5)
2 the group that follows the ideas of Darwin, Wallace and Matthew equates the centre of origin with the centre of diversity (criterion 4) and holds the opposite idea that the most evolved species live in the centre of origin (criteria 2 and 6), forcing the more primitive, less competitive forms to move away.

There is a paradox in the ideas held by the second group (Rosen, personal communication, January 2001). If new species in the centre always force out older ones, how would this generate a higher diversity in the centre than in peripheral regions? At least some of the older species would have to have remained in the centre in order for a greater diversity to build up there. Alternatively, one would have to propose that evolutionary rates were forever accelerating in the central origin, or extinction rates in peripheral areas were always greater than speciation rates in central areas.

If the areas in which multiple species are supposed to have originated are all shown to be identical, a 'centre' model would be applicable. However, the concept of 'centre of origin' is, for many authors, a place where numerous species are supposed to have originated over long periods of time. This idea of 'centre of origin' does not seem to have a practical reality.

3.2.1 Dispersal is a pattern and not a process

After summarizing the principal ideas about centre of origin, it is not surprising that the general approach of dispersal biogeography is often criticized for proposing *ad hoc* explanations, or 'just so stories' (Eldredge, 1981), for dispersal patterns of different taxa.

I would like to note that the assumption of dispersal from a centre of origin cannot be seen as a process itself because it corresponds to the description of, say, an observed pattern implying a theoretical causal sequence (dispersion → settlement → diversification in the new areas) that involves different processes. The meaning of the term 'dispersal' is unclear because it does not identify a process. It appears in fact to be an 'explanation of the pattern in terms of ideas about process' (Eldredge, 1981).

The location of the centre of origin is the goal of most studies of historical biogeography. It is certain that a given clade originates in a given place but, as discussed above, the identification of that place does not follow a single criterion or a group of congruent criteria, thus revealing numerous contradictions inherent in the 'ideas' of centre of origin.

The existence of disjunct distributions, for example taxa having relatives in areas geographically widely separated by barriers such as oceans, deserts or mountains, is problematic. The dispersal hypothesis explains such disjunct distribution as the result of successful dispersal involving extension of the distributional area from the centre of origin, the successful crossing of barriers and also the possible eventual diversification by allopatric speciation in the new habitats. But ambiguities derive from the lack of both distinction between dispersal and dispersion and precision about the time-scale at which all the abovementioned processes (except allopatric speciation) operate. Vicariists, in contrast, consider disjunct endemics as relicts of a previous, more widespread, population. Dispersalist hypotheses are not only proposed by specialists of terrestrial organisms but also by zoologists and palaeontologists dealing with planktonic taxa (van der Spoel, 1983).

3.2.2 Dispersal and allopatric speciation

The most dramatic problem encountered in the attempts to apply the centre of origin ideas concerns the speciation processes underlying the biogeographical distributions. This aspect is never very clear in the different biogeographic (and palaeogeographic) papers advocating dispersal hypotheses, therefore numerous questions arise: When and where speciation occurred? Did dispersal (i.e. long-term distributional change) occur before or after speciation? Can speciation occur after dispersion (the short-term process defined in Chapter 2)? The advocates of dispersal biogeography, as Briggs (1981) stressed, consider that allopatric speciation can result from successful migrations of organisms (from centres of origin) across geographic barriers, whereas vicariists believe that allopatric speciation takes place as the result of barriers which develop to separate portions of previously continuous populations.

Cracraft (1980) stated that palaeontology had little or no relevance to the examination of dispersal hypotheses. Vicariance explanations are considered more economical than those of dispersal. Cracraft also recommended investigations through cladistic analysis as a test for assumptions of dispersal events from a centre of origin prior to vicariance. It must be noted that he considered vicariance as a process coinciding with the process of allopatric speciation and did not distinguish between long-term dispersal (in this book called 'faunal spread' or 'colonization' if eventual speciation occurs) and short-term dispersion (in this book called 'dispersion'). Despite Patterson's (1983, p. 12) statements, 'cladistic analysis of species demands no prior assumption about speciation or even about evolution' and 'cladistic biogeography need concern pattern alone', vicariant interpretations are 'tuned' in, if not really constrained by, assuming allopatric speciation. Therefore in this respect the consequence of the separation of a previously continuous population or even biota cannot have other outcomes than the passive '*in situ*' evolution in each one of the areas resting on the opposite sides of a newly formed barrier. Lieberman (2000, pp. 77, 91) considers speciation via vicariance as a passive phenomenon preceded by the range expansion of the ancestral species. This involves an idea of centre of origin.

Communities of oceanic islands, which are separated from mainlands by wide oceanic hiatuses, are used as a classical example of the dispersal model. They possess endemic faunas and floras whose ancestors are believed to have arrived by long-distance dispersal events, sometimes passively by anemochorian and zoochorian transports, or by water currents, rafting on floating trunks (these cases are considered examples of dispersion in this book). Whittaker (1998) has summarized and discussed different examples of island colonization through dispersions of different kinds.

Provided that a viable population has been established, once organisms arrived in these islands, they would diversify and speciate if they found suitable, 'empty', environments. Then they would colonize these environments in the absence of competition. Some of these 'chance dispersal' events have recently been observed, for example the arrival at Anguilla Island (in the Caribbean) of green iguanas (*Iguana iguana*) on a mat of logs and uprooted trees after hurricanes reported by Censky *et al.* (1998). Males and females invaded the island, survived and a female in reproductive condition was observed. The authors stress that the likelihood of reproduction is high and interpreted this event 'as over-water dispersal'. They also consider high the probability of successful colonization of the island by *I. iguana*. In this example the speciation process has not yet occurred but if the predictions of Censky *et al.* (1998) were confirmed it would occur **after** a short-term event and undoubtedly would correspond to allopatric speciation. In contrast, it is demonstrated that long larval transport of some marine organisms does not lead to speciation because the genetic flow can be maintained (see Chapter 6).

3.2.3 Dispersal routes

In palaeobiogeography most researches have been devoted to identifying both the centre of origin and dispersal routes (often called migration routes). Simpson (1940) roughly classed dispersal routes of terrestrial organisms on the basis of their effectiveness in permitting organisms to pass. His terminology is still used and is also applicable to marine organisms. **Corridors** provide a high probability of dispersion from a region to another. The straits of Gibraltar allow faunal communication between Atlantic Ocean and Mediterranean Sea. The Arctic shelf is a corridor for the the circum-Arctic dispersion of benthic organisms. **Filters** allow a relatively easy dispersion to only some organisms. For example, ocean basins separating oceanic islands may be filters to some benthic invertebrates whilst they are crossed easily by species characterized by a planktotrophic larval stage (see Chapter 6). **Sweepstakes routes** permit the transit of nearly all species, although these routes are unstable and ephemeral in geological time. The term **island hopping** was used by Simpson (1962) to indicate a form of chance dispersal along sweepstakes routes. **Stepping stones** (MacArthur and

Wilson, 1967) are intermediate islands permitting the transit of faunas towards a previously uncolonized (e.g. new) island.

3.3 Panbiogeography

This is the biogeographic method introduced by Leon Croizat, involving drawing lines on a map which join the known distributions of related taxa in different areas: these lines have been called **tracks**. The course of some important tracks on an intercontinental scale is controlled by the **nodes**, i.e. areas where tracks intersect. Croizat (1952, 1958, 1962–1964) has drawn tracks of different unrelated taxa, both animals and plants, to obtain a global scale biogeographic frame that enabled him to generate geological hypotheses. His **panbiogeography** was based in fact on the principle that Earth and life have evolved together. Croizat realized that the course of some individual tracks runs parallel to others: a set of different coinciding individual tracks form a so-called **generalized track**. This means that biotic distributional patterns are non-random and fall into a limited number of tracks. The tracks for terrestrial organisms may cross oceans and hence could not be explained by present-day geography. Therefore, according to Croizat, the generalized tracks indicate that the modern disjunct patterns are very unlikely to be the result of coincidental episodes of long-range dispersal. On the contrary, they represent an ancestral biota that became vicariated into two descendant biotas as a result of some geological or climatic event.

Croizat's ideas were considerably strengthened by the widespread acceptance of the theory of plate tectonics. The rifting and drifting of continental masses provided a mechanism for the isolation of vicariant populations, thus dismissing the ancient hypotheses of dispersal on continental bridges across the oceans to explain the disjunct distributions of terrestrial organisms.

Croizat's concepts were initially supported by the cladistic biogeographers who in the 1970s formed the school of 'vicariance biogeography'. Area cladograms were compared to tracks and the sum of the areas encompassed by one cladogram resulted in the equivalent of a track. The difference between these two approaches (tracks and cladograms) is that Croizat's analytical method is essentially qualitative and has been perceived as not rigorous by some cladistic biogeographers. But Page (1987) developed a quantitative approach to Croizat's panbiogeography equating the track with the graph-theoretic concept of the minimal-spanning tree. Page represented tracks by two different matrices that enable incongruent tracks to be detected and shared elements (biogeographic homologies) to be identified. Then tracks can be oriented using both phylogenetic and biogeographic criteria. The statistical significance of track concordance can be evaluated using permutation tests of association between matrices.

As far as I know, Croizat's panbiogeography and Page's method have not been applied to palaeobiogeographic studies of marine organisms. However, the importance and the impact of panbiogeography on biogeographic thought cannot be ignored (Craw, 1988).

3.4 The vicariance model

Vicariant taxa are phylogenetically closely related taxa that occur in different areas. This concept clearly recalls that of disjunct distribution. The main historical explanations of the disjunct distributions are essentially:

1 long-range 'dispersal' (whose temporal scale is rarely specified though it is apparently long-term) from a previous area of occurrence to a particular site with absence of the taxon in the geographically intermediate areas
2 separation of formerly continuous populations by creation of barriers preventing any possibility of the disjunct populations mixing, or by disappearance of the taxon in the intermediate areas.

This second explanation clearly considers allopatric speciation as the main evolutionary effect of the separation of a population. The fragmentation of ancient continental masses and their drift during geological times may explain disjunctions in biotic distributions: this process (actually a summation of different short-term and long-term processes) was termed 'vicariance'. Therefore the general acceptance of the plate tectonics theory gave rise to vicariance biogeography, which grew in the 1970s as an alternative to the more traditional 'dispersalist' approaches to historical biogeography.

The results obtained from the panbiogeographic method led Leon Croizat to postulate that biotas may have moved as units attached to and in association with the movement of continental masses. The vicariance approach generally requires cladograms (explicitly formulated hypotheses of relatedness among taxa) and thus must be preceded by a phylogenetic analysis of the taxa from the region of interest. The latter method is called **cladistic biogeography** (see Chapter 5). Although some vicariance biogeographers do not use cladistic techniques (see McCoy and Heck, 1983), and also Croizat's method was not cladistic (Croizat, 1982), it is probably the most used analytical method of vicariance biogeography. In fact, branching phylogenetic patterns in a taxonomic group can be interpreted in terms of history of vicariance events because taxonomic divergence is assumed to be the result of the subdivision of a previously unified area.

While acknowledging that 'dispersal' (short-time, long-time or both?) can take place, vicariance biogeographers do not consider that successful crossing of barriers by dispersal is an important process in establishing world biogeographic patterns (Cracraft, 1980; McCoy and Heck, 1983; Wiley, 1988). Most species are considered to have evolved *in situ* by subdivision of the parent population rather than by dispersal, thus vicariance appears the most parsimonious assumption and, furthermore, has more credibility as it is rationally linked to Earth process. Therefore disjunct distributions, the real playground of the competition between vicariance and dispersal hypotheses, are the result of range reduction rather than expansion and, in this sense, vicariance appears as a 'passive' process. The areas of origin are 'regarded as relatively unimportant and perhaps unknowable' (Patterson, 1983, p. 14); although their possible existence is not denied, the question 'becomes unnecessary and perhaps unknowable' (Platnick, 1981, p. 149).

Historical processes leading to disjunct distributions of some taxa are usually explained by dispersal biogeographers as successful crossings of barriers which are considered by vicariance biogeographers to be the result of chance that can be neither predicted nor verified. The distributions of planktonic and nectonic organisms are often explained through dispersalist hypotheses supported by distributional data of areas of high abundance and diversity which are usually interpreted as centres of origin. The disruption of a genetic continuum with extinction of the possible ancestor group in intervening areas and subsequent speciation has been invoked by van der Spoel (1983) to explain the disjunct distribution of pelagic taxa (shrimps, pteropods) on the opposite sides of the Pacific Ocean. The distribution of planktonic organisms in the so-called Central-Water masses in the Atlantic, Indian and Pacific Oceans seems to be the result of disruption of an original area, combined with extinction and adaptations without expansion, i.e. vicariance, rather than recent dispersal.

Certainly larval transport, particularly the dispersion of planktotrophic teleplanic larvae, can result in a successful crossing of barriers, as shown by the distribution in the American West Coasts (Pacific) of living Indo-West Pacific tropical molluscs beyond the Great Pacific Barrier (Newton, 1988). But, as already noted, this process corresponds to the maintenance of the genetic flow. Geographic models of speciation that rely on limited dispersal capability as a means of creating genetic isolation and divergence, such as vicariant and allopatric hypotheses, have been considered not as satisfactory for explaining pelagic biodiversity as are sympatric and parapatric models (Norris, 2000). This author suggested that speciation in pelagic taxa may occur by a change in the place of reproduction, for example water depth (depth parapatry or sympatry) or by a change in timing of reproduction (seasonal sympatry or parapatry). It is worth noting that Norris' interpretations contrast with Schneider and Kennet's (1999) parapatric and allopatric models of depth segregation and speciation for the planktonic foraminifers of the genus *Globoconella* (Fig. 3.1).

Vicariant patterns are clearly the result of long-term processes. Dispersion of planktotrophic larvae is a short-term process that maintains the gene flow between patchily distributed populations of the same species across oceanic barriers, thus preventing allopatric speciation. Dispersion can even produce the expansion of the range of a species by means of non-reproductive pseudopopulations (see Chapter 6). Therefore, at least for marine invertebrates characterized by planktotrophic larval stages or other modes of dispersion (e.g. cephalopod's egg dispersion prior to hatching), it is difficult to imagine a process of allopatric speciation resulting 'from successful migrations of organisms across geographic barriers' (Briggs, 1981, p. 306) that, on the contrary, prevent speciation.

A possible reason for the difficulties of some marine biogeographers to accept vicariist interpretations may be the relative intangibility of barriers that in the sea would isolate marine populations.

3.4.1 Vicariance models in marine palaeobiogeography

Because in most marine invertebrates the maintenance of the gene flow depends on larval characteristics (though it is not always the larval stage that is the vagile one), a possible barrier is the **critical dispersion distance** (Rosen, 1984). This is conceived as a fixed attribute of a given species, i.e. its maximum (probable) potential dispersion distance. It should not be intended as a linear distance because of variable paths of current flows. Factors such as plates moving and sea-level fluctuations (which may in turn provoke coastal displacements, changes in oceanic circulation and climatic changes) may increase the distance between, say, two islands beyond the critical dispersion distance, thus leading to allopatric speciation.

Rosen (1984) has explained the notable diversification of reef corals in the Indo-Pacific area during the Late Cenozoic through allopatric speciation. Starting from the concepts of critical dispersion distance and **staging posts** (Rosen, 1983), this author proposed a vicariance model called **island clustering**.

3.4.1.1 Staging posts

The distribution of reef organisms depends on their larval characteristics, so the distributional range of a reefal species must depend on its critical dispersion distance. 'Islands which lie within this limit can serve as staging posts between other islands which are otherwise out of larval reach from each other' (Rosen, 1983, p. 5). This concept is preferred to that of 'stepping stones' because that has 'overtones of chance dispersal' (ibid.) and is more concerned with first colonization of islands. Here it is the maintenance of gene flow and steady state dispersion that is concerned. Darwin's idea of atoll formation around subsiding basaltic volcanic formations has been demonstrated. The subsidence is related to sea-floor spreading, which is characterized by both lateral and vertical components. The tectonic motion of oceanic seamounts, which are often capped by reefs, affects the pattern of staging posts across oceanic regions. Allopatric speciation could result in reefal populations becoming divided by the disruption of the staging post pattern and, conversely, new staging post patterns can merge the ranges of pre-existing species. The different tectonic histories of the different islands and the relative horizontal motion between them may change staging post patterns. Convergent motion bring islands closer together, with consequent merging of marine and terrestrial biotas. This phenomenon has been called 'island integration' by Rotondo *et al.* (1981) and 'Noah's Ark' (McKenna, 1973) on a continental scale (for example India relative to Asia). Divergent plate motions increase the staging post distances until these exceed the limits of the critical dispersion distance (e.g. islands on opposite sides of the Atlantic), thus representing a potential vicariant factor. A population of reef corals which occupies an island or a patch is also linked, in population terms, with other such islands or patches by the mobile larval populations circulating between them; the subpopulations of the different patches constitute a **metapopulation** (see Chapter 6). Rosen cautioned that effectiveness and distribution of staging posts can be affected by climatic change on ocean circulation and sea-level changes.

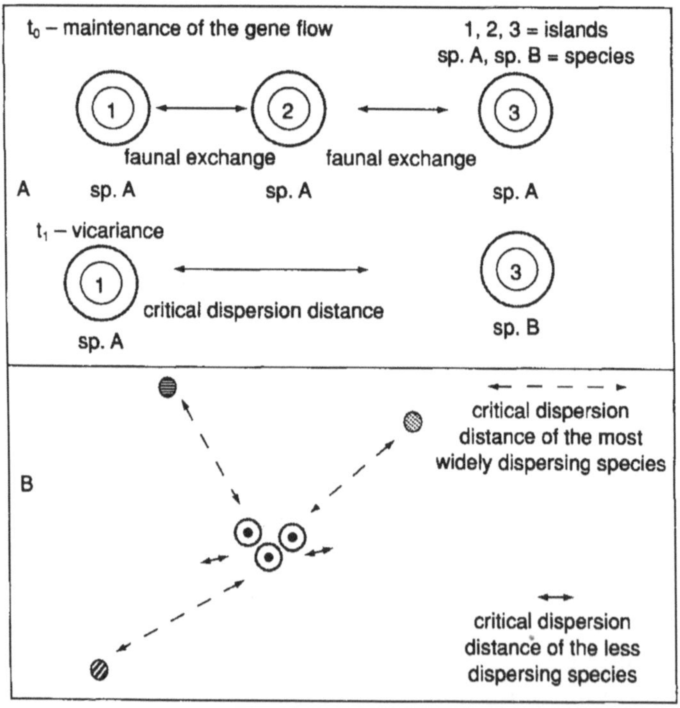

Figure 3.2 Author's schematic illustration of the island clustering model of Rosen (1984). A, three reefal islands share at t_0 the same meta-population but at t_1 two different species originated by vicariance after the drowning of island 2, a former staging post (but note that changes in currents could have similar sequences); B, two systems of three islands whose faunal content, represented by the different shading patterns in the different islands, depends on the different critical dispersion distances of the species.

3.4.1.2 Island clustering

Consider three reefal islands at time $\mathbf{t_0}$ (Fig. 3.2A): island 3 is too far (beyond the critical dispersion distance) from 1 but 2 is intermediate, thus of crucial importance for the maintenance of gene flow between the subpopulations on the other two (1 and 3). Island 2 therefore plays the role of staging post. If at time $\mathbf{t_1}$ island 2 is drowned, the sub-populations of islands 1 and 3 remain isolated because their distance is greater than the critical dispersion distance. Probability of vicariance is enhanced in this case. Consider now two systems of three islands

(Fig. 3.2B), both within the same faunal region with the same potential fauna from the regional pool. The closer set of islands are more likely to fall within the critical dispersion distances of all the available species. The scattered set of islands will fall within the dispersion distances of only the most widely dispersing species. A few short-range species will be endemic to one or other of the scattered set but will not occur in all three of them. Clustered groups of islands will carry more species than scattered ones. The more widely scattered a set of islands, the more likely it is that allopatric speciation will occur. However, this likelihood will also begin to decrease again with increased scattering of islands, if their typical distances apart are far in excess of critical dispersion distances. Conversely, very closely grouped islands are likely to carry (even accumulate) more species but are less likely to be areas of vicariance because distances of separation of islands are well within most critical dispersion distances.

3.4.2 Bipolarity: do climatic changes promote vicariance?

One of the most important features of large-scale biogeography is the pattern of disjunct distribution known as **bipolarity**. An array of explanatory hypotheses exists in the literature: Crame (1992) has summarized the most important dispersalist and vicariant views. Dispersalists postulate either that taxa arise in the Tropics and subsequently migrate to high latitudes or the opposite, with dispersal from the poles towards lower latitudes. Crame stresses that these explanations are hardly testable and cannot be falsified in that they essentially reside on narratives that imply a putative centre of origin. The vicariant explanation assumes that polar taxa represent the remnants of former cosmopolitan distributions. The fragmentation of the range of a formerly widespread biota could have been realized through tectonic fracturing of a north–south trending continent in the equatorial regions, as for example the break-up of Pangaea during the Late Triassic–Early Jurassic interval (Marcoux *et al.*, 1993). But Crame (1993) reached the conclusion that climatic changes also enhanced vicariance events of bipolar distribution splitting an ancestral biota. Widespread or cosmopolitan distributions are held to have formed during global cool phases (for example, the interval Late Eocene–Early Miocene), only to be disrupted by global warming (e.g. late Early Miocene) causing expulsion of the ancestral stenothermal-cool biota that formerly stretched continuously from northern to southern high latitudes. Crame's fascinating hypothesis is supported by two important lines of evidence:

1 climatic changes are now demonstrated to have occurred frequently in geological history
2 the geological record shows repeated distributional changes resulting in expansions and contractions of the tropical biota which suggest climatic change.

Crame (1993) identifies three main phases of bipolarity (but minor phases are also suggested): a Jurassic–Cretaceous one linked to the disintegration of Pangaea; a Late Palaeogene–Early Neogene one based on climatic change; and a Plio-Pleistocene–Recent one based on glacial cooling. Benthic molluscan groups offer considerable potential for the investigation of the Meso-Cenozoic bipolarity. Bivalves show clear bipolarity since at least the Early Jurassic, as demonstrated by the distribution of a variety of pectinaceans during the Pliensbachian (Crame, 1992, 1993; Damborenea, 1993, 1998). Another phase of bipolarity is evident in the Tithonian when inoceramid, buchiid and oxytomid bivalve occurrences at northern hemisphere high latitudes are matched with those in southern hemisphere high latitudes. A strong debate exists about the existence of bipolarity during the Jurassic (summarized by Enay and Cariou (1997) and Westermann (2000b)). In fact, a Boreal Realm is recognized on the basis of both bivalves and ammonites in the Pliensbachian and in the Late Bajocian–Tithonian interval whereas an Austral Realm seems to be clearly established since the Tithonian (Enay, 1972). One of the difficulties of recognizing a complete bipolarity in the Jurassic probably resides in the latitudinal interval of the intervening Tethyan Realm, which asymmetrically stretches from tropical latitudes in the northern hemisphere up to high latitudes in the southern one (see Fig. 2.4).

Dispersalist explanations of bipolar distribution are not rejected outright by Crame, who recalls that some gastropods characterized by considerable depth tolerances may have dispersed from one polar region to the other, crossing the Tropics by isothermal submergence. Vicariant or dispersalist hypotheses can be tested on the basis of characters of both gastropod protoconchs and bivalve prodissoconchs which allow dispersive or non-dispersive larval development to be recognized (see Chapter 6).

3.5 Centres of origin in palaeobiogeography

Analysis of the distribution of marine extant and fossil faunas demonstrates that it is possible to delimit areas of high diversity containing biotas characteristic of that area and distinct from biotas of other areas. Because many of the components of a single biota share a common history, the simple explanatory model is that of the 'centre of origin', which assumes that taxonomic richness has resulted from a concentration of evolutionary novelties over tens of millions of years. Classic examples are the reef coral fauna of the Indo-West Pacific area compared with the Atlantic one (no species and only 7% of genera in common) and the Tethyan and Boreal Realms in the Jurassic.

The idea of 'centre of origin' implicitly needs stability of the general physical and environmental conditions over geological time and implies range expansions of taxa moving in a stable geographic background. These assumptions can no longer be accepted (McCoy and Heck, 1976) due to the dramatic implications of increasing evidence of global climatic changes, recurrent changes of the general regime of oceanic currents and plate tectonics. The last of these leads to geographic changes that modify climate and oceanic circulation.

In the past three decades the classical, highly contradictory, concept of 'centre of origin' has been reconsidered in various ways. There is empirical evidence of biodiversity peaking in the Tropics and decreasing polewards as well as evidence that steepest latitudinal gradients of biodiversity are associated with the youngest clades (see Chapter 7). An alternative model to the idea of '**cradle**' (centre of origin) is that of '**museum**' (centre of accumulation, evolutionary sinks). The dichotomy 'cradle or museum' arises from a famous question posed by Stebbins (1974) for Tropics as a whole. A third model is the '**refuge**' (centre of survival), i.e. a suitable place for an organism to live, into which there is evidence that the range of the organism has contracted (Rosen, 1984). While accepting the possibility of dispersal events in some cases, the 'museum' and 'refuge' models mainly focus on vicariance to explain allopatric speciation.

An analysis of the myriad biogeographic studies is impossible here, thus I have selected three examples for discussion.

3.5.1 Is the Indo-West Pacific Focus a centre of origin, accumulation or survival?

One of the most celebrated 'centres of origin' is the Indo-West Pacific (IWP) area, whose waters contain the highest marine faunal diversity in the world (land faunas are also among the richest). Biogeographic studies of reef corals have fuelled different hypotheses, which have potential implications for other groups and for tropical reef ecosystems as a whole as coral reefs provide habitats for many marine organims. The IWP Focus has been considered:

1 a centre of evolutionary radiation (Ekman, 1953; Stehli and Wells, 1971; Briggs, 1974; Veron, 1995), or centre of origin, where conditions have been favourable during a long time to generate new taxa continually there
2 a centre of survival, or refuge, where the modern fauna would correspond to both a vestige of a formerly rich, cosmopolitan fauna of Cretaceous origin and the recent accumulation of new taxa (McCoy and Heck, 1976)
3 a centre of accumulation consisting of species originated in other areas than IWP, whose geographical ranges changed over time to coincide in the present IWP high-diversity Focus (Newell, 1971; Rosen, 1984, 1988c; Pandolfi, 1992).

Wilson and Rosen (1998) have summarized these different models in a scheme which is reproduced in Table 3.1.

Stehli and Wells (1971) derived different patterns from the distributional data of modern reef corals and also used the fossil record to obtain age-and-area patterns as a further support to the idea of the centre of origin. These authors demonstrated that coral richness progressively declines from the very centre of the focus to its periphery in all directions, and that stratigraphically youngest genera are restricted to the IWP Focus, whereas stratigraphically older genera are more widespread (Fig. 3.3).

These data for hermatypic corals have been interpreted 'to suggest that centres of diversity are also evolutionary centers where new forms evolve and from which they spread to marginal areas' (Stehli and Wells, 1971, p. 119). However, Stehli and Wells' patterns have been inferred from data which are, admittedly, means of all stratigraphical ages for each station of sampling. Present-day distributions cannot, of themselves, tell us where an organism originated. Age patterns in IWP reef corals are the result of different events or processes which cannot be predicted by age-and-area 'rules' alone. Widespread taxa generally have long stratigraphic ranges and the converse also applies, but some old reef coral genera have narrow distributions and the younger genera are not more randomly distributed than the older ones. As Boucot (1975b) has shown, the older a taxon, the less clearly related is its latest known distribution to its earlier history, especially its 'origin' (see Fig. 2.7).

McCoy and Heck (1976) explained the high diversity in the IWP area as a result of an interplay among speciation, extinction and tectonic events. The biota would be represented by both old groups, a remnant of a formerly widespread biota, and recent taxa accumulating

Table 3.1 The three main models for the origin of the IWP Focus. The main evolutionary characteristics within and beyond the centre, together with predicted patterns for fossil record, are shown. 'Oldest records' refers to the different criteria about the age and origin of the taxa within the central region commonly considered to determine the centre of origin (Wilson and Rosen, 1998).

		'CRADLE' source: centre of origin	*'MUSEUM' sink: centre of accumulation*	*'REFUGE': centre of survival*
WITHIN THE FOCAL REGION	**Speciation rates**	high	low	low
	Extinction rates	low	low	low
	Immigration rates	low	high	low
OUTSIDE THE FOCAL REGION	**Speciation rates**	low	low	low
	Extinction rates	high	high	high
	Immigration rates	high	low	low
PREDICTED DIVERSITY PATTERNS THROUGH GEOLOGICAL TIME IN FOSSIL RECORD		Diversity conspicuously higher in the centre than in outside regions, and of greater duration than species turnover time-scale (<30 Ma)	Increasing high diversity through time in the centre, relative to outside regions	No particular pattern of change in high diversity in the centre, through time, but diversity elsewhere diminishing through time
OLDEST RECORD		Mostly within centre	No particular pattern or mostly outside centre	No particular pattern

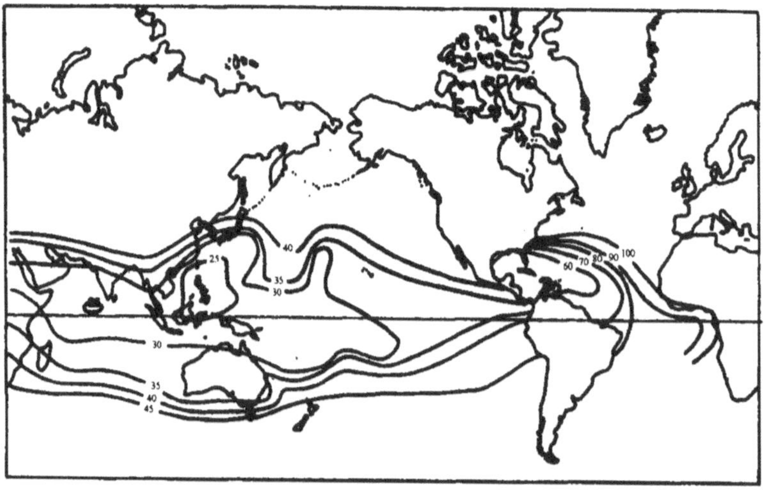

Figure 3.3 Average generic age data for hermatypic corals.

Source: Stehli and Wells (1971). Modified with permission of the editor.

through colonization of the numerous islands which provide greater shoreline area and increased chances for isolation and diversification. These explanations were considered not necessarily inconsistent with a centre of origin concept, though originations in the IWP with subsequent outward dispersal on the one hand, and repeated long-distance dispersals to explain the disjunct distributions of some taxa shared between the Caribbean and the IWP areas on the other, are considered very infrequent or unlikely.

The succession of historical events, biotic and abiotic, provide causal explanations of the modern distributional patterns. Accordingly, the notable diversification of reef coral taxa in the IWP during the Cenozoic has been explained by Rosen (1984, 1988c) by vicariance processes induced by Neogene glacio-eustatic fluctuations and accumulation (explained by the model of island clustering) of different isolates as a result of both the northward migration of the Australian plate and the associated tectonic processes which created new land masses in the equatorial region between the SE Asia mainland and Australia. Islands may have functioned as staging posts in a first phase and the northward migration of the Australian plate would have accumulated the different isolates in the IWP area in a second phase. The historical contingencies considered by Rosen do not support the model of the

IWP Focus as a 'centre of origin', which should be recognizable in the fossil record as a region of persistent high diversity. However, Wilson and Rosen (1998) have definitively demonstrated that the patterns observed by Stehli and Wells (1971) are strongly influenced by the Late Cenozoic signal because the IWP Palaeogene fauna is extremely poor (the 'Palaeogene gap'). The IWP high diversity seems therefore a geologically young, probably Early Miocene, feature. The recognized cool phases cannot explain the 'Paleogene gap' since the western part of SE Asia roughly remained in tropical latitudes during the Cenozoic. Geographical isolation from other coral-rich regions, relative lack of habitable areas, and local ecological factors are considered key factors. Neogene collisions of Australian fragments with SE Asia led to decreased isolation and formation of numerous shallow-water areas with diverse and abundant reef corals in the Miocene. The geographical complexity of IWP area may well then have favoured localized isolation and origination of new taxa, but the fauna is also made of older relicts and taxa which originated elsewhere.

A younger age has been suggested by Budd *et al.* (1994) for the Caribbean reef-coral fauna, which achieved a distinctively modern aspect (>60% generic similarity) in the Early to Middle Pleistocene. Furthermore, some coral genera which are among the most important constituents of the modern eastern Pacific fauna have been recognized to have become extinct in the Caribbean during the Neogene. This suggests that the dispersal hypothesis for the origin of the eastern Pacific fauna may be oversimplified. The new fossil data analysed by these authors also suggest that both the eastern Pacific and Brazilian faunas may partially represent relicts of the more generically diverse Neogene Caribbean reef-coral fauna.

The Modern latitudinal biodiversity gradients, in reef corals at least, are partly the result of historical contingencies and do not favour the idea of the Tropics as 'a cradle' for a substantial period of geological time. On the other hand, tropical biodiversity could have been increased by concentration of molluscan taxa resulting from climatic changes (Crame, 1996). Noteworthy in Rosen's hypotheses is the historical reconstruction based on both the fossil record and the geological data, that is lines of evidence which are independent from the present-day distributional patterns of reef corals. The combined patterns from a range of different studies led Wilson and Rosen (1998) to conclude that a more appropriate model requires a combination of all three of the possibilities listed in Table 3.1. Plate tectonics processes have probably been a major control on IWP diversity patterns of reef corals and presumably also other shallow marine organisms.

3.5.2 The bathyal fauna of the South-West Pacific seamounts

The deep-sea environments are characterized by highly diverse faunas. Richer de Forges *et al.* (2000) have recognized highly diverse bathyal benthic faunas characterized by numerous endemic species (even though deep organisms are generally considered pandemic) and relict forms in the South-West (SW) Pacific seamounts (on the Lord Howe Rise off eastern Australia, Norfolk Ridge off South-East New Caledonia and off southern Tasmania). It is worth mentioning here the most important archaic elements of the fauna.

Sphinctozoans (calcified sponges), a group considered to have become extinct at the end of the Cretaceous, have been reported from Norfolk Ridge mainly, but also all around New Caledonia (Vacelet *et al.*, 1992); the living representatives belong to the Triassic family Cryptocoeliidae. A living inarticulate brachiopod of the family Craniidae, *Neoancistrocrania norfolki*, was described by Laurin (1992) again from Norfolk Ridge; its closer relative is the Cretaceous genus *Ancistrocrania*, which was widespread in the Northern Hemisphere. The stalked crinoid fauna reported from bathyal environments off north-west and south-east New Caledonia by Améziane-Cominardi *et al.* (1987) appears to be the most archaic in recent oceans, showing close relationships with fossil faunas of the Mesozoic Tethys. Among the archaic taxa the authors reported the only living representative of the Palaeozoic Subclass Inadunata (*Guillecrinus*, a genus also reported from Ile de la Réunion, western Indian ocean), *Proisocrinus ruberrimus*, the only living representative of the Jurassic family Millericrinidae, as well as species of the family Bathycrinidae showing their closest affinities with Late Cretaceous forms.

Endemicity of benthic faunas living on relatively close seamount ridge systems (e.g. Lord Howe Rise and Norfolk Ridge) appears to be greater than that of deep-sea vent communities of disjunct ridge systems which, on the contrary, display a relative similarity. Richer de Forges *et al.* (2000) explain these features arguing that vents are relatively ephemeral environments, whereas seamounts are typically millions to tens of millions of years in age. Therefore, dispersion between vent sites appears to be essential to the continuance of these communities whereas taxa adapted to seamount conditions limit dispersion to maintain their populations, because of the generally small size of seamounts.

The high diversity of the bathyal benthic seamount faunas, their endemicity and the occurrence of living representatives of archaic forms could be suggestive of a centre of origin located in SW Pacific if one refers to criteria listed in Table 3.1, in particular high speciation rates and low extinction rates. However, there is a strong conflict with the six criteria listed in section 3.2.

Améziane-Cominardi *et al.* (1987) stressed that both the rate of archaism and the distributional patterns of the stalked crinoid fauna, characterized by both pandemic and endemic elements, do not fit the model of centre of origin. This fauna derives from the Mesozoic Tethyan fauna, which was separated in three assemblages (located in the Atlantic, West Pacific and Indian oceans) after the closure of the Tethys ocean and the formation of the apino-himalayan chains. However, the highest taxonomic diversity of western Pacific stalked crinoids is observed in an area from Sunda Islands to the southern Sea of China. Outside this area the trend of diversity decrease is paralleled by the maintenance of the archaic taxa.

After the disruption of the ancestral Tethyan fauna, the colonization of the oceanic floors by stalked crinoids took place along the oceanic ridges (Roux, 1982). The closure of the Tethys generated a sort of 'museum' effect in West Pacific but formation of new ridges provided new routes (which can be described as corridors) for colonization in the Pacific, also generating vicariant events, as suggested by low similarity between faunas of Norfolk Ridge and Lord Howe Rise. Distinct plate tectonics events concurred to maintain the ancestral taxa on the one hand and to favour the colonization of both oceanic floors and seamounts on the other hand.

The history of this bathyal fauna appears to have been quite complex and cannot be simplified by postulating centres of any kind.

3.5.3 Tethyan and Boreal Realms during the Jurassic and Cretaceous

I must resort once more to the Mesozoic and particularly to ammonites and related pelagic communities intensively studied in Europe for more than 150 years. After the crisis of the end of the Triassic, a new, spectacular, ammonoid radiation occurred. Ammonites occupied different environments from the epioceanic Tethyan areas to intracratonic basins and shallow epicontinental seas. Two centres of origin, Tethyan and Boreal, commonly quoted since Arkell (1956) as realms, i.e. inclusive high-rank biogeographic units, are classically distinguished by palaeobiogeographers (see the recent review by Westermann (2000b)). The boundaries of both the Tethyan and Boreal Realms, which corresponded to distinct biotas, continuously fluctuated (see Fig. 4.3) under the control of eustatic, climatic and tectonic factors (Enay, 1980b; Hallam, 1983; Callomon, 1985; Hoedemaeker, 1990; Rawson, 1993).

In the cratonic seas of Central and Western Europe, well-documented range expansions and contractions of the two faunal realms cause the repeated shifts of the biogeographical boundaries (Fig. 3.4). The resulting dynamic interactions between the two biotas led to faunal turnovers and major evolutionary events among ammonites (Enay, 1980b; Enay and Mangold, 1982; Dommergues and Marchand, 1988; Oloriz *et al.*, 1993; Rawson, 1993).

Tethyan-Boreal provinciality is also shown by belemnites, brachiopods and forams but is difficult to recognize in bivalves (Hallam, 1994) despite the fact that characteristically Tethyan elements exist (*Lithiotis*, rudists, etc.). Calpionellids were endemic in the Tethyan Realm (Enay, 1972; Remane, 1998). Since Neumayr and Uhlig (and more recently Ziegler (1974)), these opposite biotas have been considered as the expression of a temperature-controlled, roughly north–south, latitudinal zonation. However, modern palaeogeographic reconstructions based on plate tectonics (Dercourt *et al.*, 1993; Ricou, 1994) show a more complex picture for the latitudinal distribution of both the Tethys, i.e. the palaeogeographic oceanic domain, and the biogeographic Tethyan Realm itself (Cecca, 1999). Therefore, temperature does not appear to be the sole factor of provinciality between and within Tethyan and Boreal Realms.

Biogeographically, a Boreal Realm can be distinguished in the Pliensbachian, when it is essentially characterized by the ammonite families Amaltheidae and Liparoceratidae, and from the Late Bajocian onwards. The intervening time interval from Toarcian to Bajocian is generally considered to be characterized by a relative cosmopolitanism (though provincial distributions are recognized at the genus level). The distinction of the Boreal and the Tethyan Realms in the Pliensbachian is not based simply on the birth and the distribution of the families Amaltheidae and Liparoceratidae but also on a phase of endemicity paralleled by a morphological ammonite diversification on the different habitats (Meister and Stampfli, 2000; Dommergues *et al.*, 2001).

Both realms have been treated, more or less explicitly, as 'centres of origin'. The Tethyan Realm was persistently characterized by higher diversity, though belemnite generic diversity increased towards the northern, Arctic, areas of the Boreal Realm (Doyle, 1987, 1992). Authors either avoid expressing their ideas about speciation or have gradualist explanations supported by dispersalist scenarios. Vicariance has been rarely proposed (but see Dommergues, 1987; Dommergues and Marchand, 1988; Cecca, 1999).

However, in agreement with classical views postulating the origination of taxa in the Tropics followed by dispersion at higher latitudes, the

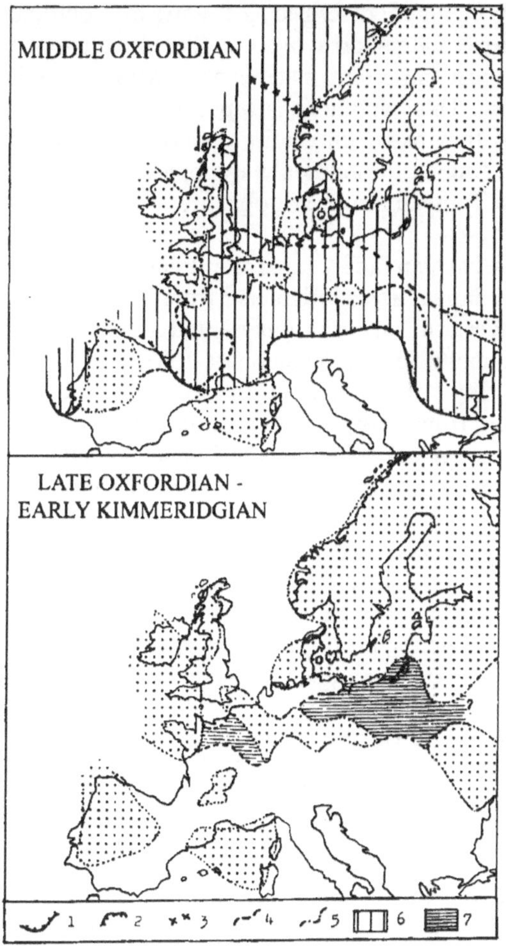

Figure 3.4 Shifts of the biogeographical boundaries of Boreal and Tethyan ammonite faunas in North-Western and Central Europe during the Middle Oxfordian–Early Kimmeridgian interval. 1, southern boundary of Boreal faunas; 2, northern boundary of Tethyan fauna; 3, northern boundary of corals during the Middle Oxfordian; 4, northern boundary of Tethyan oppeliids during the Middle Oxfordian; 5, northern boundary of the ammonite genus *Gregoryceras* during the Middle Oxfordian; 6, areas of biogeographical overlap (interface/intergradation and secondary provincialism *sensu* Dommergues and Marchand, 1988); 7, barrier formed by the 'sequanian' facies separating Boreal and Tethyan faunas during the Late Oxfordian and the Early Kimmeridgian (primary provincialism). Probable land areas are represented with dots.

Source: Enay (1980a). Modified with permission of the Société Géologique de France.

Tethyan has often been described as a centre of evolutionary radiation supplying species to the Boreal one, considered evolutionarily less effective for the most important macroevolutionary events in ammonites (Enay, 1980b). This hypothesis is supported by inferred phylogenetic relationships (not based on cladistic analysis) between Tethyan and Boreal taxa. Among the ammonite examples usually taken to support the idea of a Tethyan centre of origin are:

- the Pliensbachian family Liparoceratidae, which essentially spread in to the Boreal areas, originated from Tethyan ancestors and its North-Western European representatives gave rise to the Boreal family Amaltheidae (Dommergues, 1987)
- the typically Boreal families Cardioceratidae and Kosmoceratidae originated from Tethyan ancestors and instantaneously (at a geological time-scale) colonized the Boreal Sea after a considerable period during which it had been vacant of ammonites (Callomon, 1985)
- the family Aulacostephanidae appeared in the Late Oxfordian deriving from Tethyan perisphinctids which reached the boreal seas during the Middle Oxfordian and remained isolated there after a marine regression (Enay, 1980b)
- the family Dorsoplanitidae originated in the Late Kimmeridgian from Tethyan ancestors (Enay, 1980b; Hantzpergue, 1989)
- the subfamily Virgatitinae, which colonized the Russian platform at the end of the Jurassic, had a Tethyan ancestor (Kutek and Zeiss, 1974).

Similar evolutionary events have been described by Rawson (1993) for Early Cretaceous ammonites.

Dispersal scenarios have been suggested to explain these evolutionary events, although vicariant patterns could have resulted from the repeated isolation of Tethyan ancestors in the boreal seas. In other words, barriers would have been crossed through dispersion of propagules and eventual speciation took place after geographical isolation. However, we should note that the evolutionary events mentioned above took place **within** the Boreal Realm.

Opposite examples of Tethyan taxa with Boreal ancestors also exist, though they are far less numerous. For example, the Tethyan Valanginian family Oosterelliidae originated from the Boreal genus *Platylenticeras* (Cecca, 1998); the Tethyan Hauterivian genus *Lyticoceras* probably derives from the boreal *Endemoceras* (Rawson, 1993), although authors still discuss the possible Tethyan derivation of both *Platylenticeras* and *Endemoceras*. In any case, there is evidence too for an 'autochthonous' Boreal origin of numerous Jurassic and Cretaceous ammonoid taxa in the Boreal Realm.

It is certainly true that numerous Tethyan taxa considerably expanded their distributional ranges and also colonized new areas, but curves of ammonoid origination rates in Boreal and Tethyan areas for the whole Jurassic–Cretaceous interval have yet to be investigated following the recent example of O'Dogherty *et al.* (2000) on Jurassic ammonites from Southern Spain. The proportions of the higher diversity of ammonite faunas in Tethyan areas, compared to the Boreal ones, still wait to be assessed quantitatively. Bivalve bipolar distribution in the Jurassic is consistent with vicariant explanations (Crame, 1993) that weaken the dispersalist idea of a Tethyan centre of origin by assuming the tectonic fragmentation of a formerly pandemic biota. In the Tithonian (albeit an incomplete fossil database), bivalve and gastropod diversity shows steep gradients from the northern mid to high latitudes. The far less clear-cut diversity gradient in the southern hemisphere, a figure similar to the difference of the diversity gradients of modern bivalve faunas in the two hemispheres (Stehli *et al.*, 1967), is also one of the reasons for the debate about the existence of an Austral Realm (i.e. bipolarity) during the Jurassic (Enay and Cariou, 1997; Westermann, 2000b).

Many palaeobiogeographical maps contain arrows which in effect represent faunal spreads from centres of origin: the palaeobiogeographic and evolutionary relationships between Boreal and Tethyan Realms have been represented by authors including myself (see Cecca and Enay, 1991; Cecca, 1998) in this way. However, it is impossible to consider these supposed centres of origin as 'factories' of new taxa whose cores (not the fluctuating marginal areas) remained geographically stationary for millions of years. The repeated faunal expansions and contractions driven by climate changes and sea-level fluctuations lead me to consider that taxa did not spread from a 'centre of origin'; instead, the biogeographical range of the biota changed as a response to new environmental conditions. For example, the arrival of ammonites in shallow platform areas where previous environmental conditions were not suitable for such animals indicates changes in the volume of ammonites' ecospace. The effects of eustatic fluctuations were particularly strong on epicontinental platforms and their margins, whereas they were much less important in distal, epioceanic environments (Oloriz *et al.*, 1993; Hoedemaeker, 1995). However, O'Dogherty *et al.* (2000) suggest that faunal turnovers in Jurassic ammonites of the distal epioceanic environments of the Betic Cordillera (which was part of the

Western Mediterranean Tethys) are the expression of global changes in sea-level: extinctions and radiations would be linked respectively with sea-level falls and rises. But diversity trends may be reversed during either a single rise or a single fall of the sea level (see Chapter 7). Substantially different factors enhanced speciation not only in distal, epioceanic areas (Cecca, 1998) but also within epicontinental seas. Sea-level change was not, *per se*, a factor of speciation but it drove changes and shifts in ecospaces and environments. This process corresponds to environmental tracking. Accordingly, repeated fluctuations of the biogeographic boundaries of Tethyan and Boreal Realms do not correspond to range expansions from centres of origin but to ecostratigraphic events.

The eustatism was not the sole factor controlling the relationships between Tethyan and Boreal biotas. The repeated shifts of the biogeographical boundaries, as well as the extent of the areas of faunal overlap, were also controlled by climatic events and plate tectonics. Complex causal interrelationships between these different factors determined the distributional patterns discussed in Box 2.1. For example, the northward shift of the northernmost occurrences of both corals and Tethyan ammonites coupled with the northward shift of the southernmost limit of the Boreal ammonite faunas in the Middle Oxfordian (Enay, 1980b; Enay and Mangold, 1982) may correspond to a climatic warming. But climate does not explain all episodes of shifting of faunal boundaries. The southward shift of the northernmost limit of Tethyan ammonites in the Early Kimmeridgian may suggest a decrease in water temperature, also suggested by isotopic data (Weissert and Mohr, 1996). Accordingly, a southward expansion of Boreal ammonites is expected during a cooling phase but actually they retreated northward. As already shown by numerous studies, during both the Middle and the Late Jurassic the Boreal ammonites did not reach the distal, epioceanic environments of Tethys. Their southernmost occurrences roughly correspond to the boundary between the southern margin of the Eurasian continent, which was certainly an ecological boundary (Dommergues and Marchand, 1988). During Middle to Late Jurassic time this boundary shifted northward as plates moved (Dercourt *et al.*, 1993), thus showing that tectonics also controlled these fluctuating ammonoid distributional patterns.

In conclusion, the classical idea of Tethyan and Boreal Realms as centres of origin appears to be an oversimplification. The different evolutionary dynamics of biotas should be tested with heuristic explanations and substantiated by deep, quantitatively-based, analyses of the evolutionary patterns (e.g. origination and extinction curves). Most historical explanations of Jurassic and Cretaceous palaeobiogeography contain an underlying 'tethydism' (this term is mine), conceptually equivalent to 'olarcticism'. The latter is the combination of an array of hypotheses (derived from the ideas of Darwin, Wallace and Matthew) which tended to explain the origin of phyletically related but biogeographically disjunct clades through dispersal from centres of origin located in the Boreal hemisphere.

4 Biogeographic analyses and biogeographical classification

After the definition of the basic biogeographical concepts, numerous questions arise. How can the observed biogeographical distributions be explained? What are the causal processes? How are the major geological processes involved in shaping the biogeographic distributions? Some of these questions may remain unanswered when the aims of the biogeographical study are not precisely defined. The aims can be expressed in terms of investigative pathways which apply to both the extraction and the explanation of biogeographical patterns, to the investigation of processes and to reconstruction of evolution and palaeogeography. Because the process is inferred from the pattern, we must take into account the limits and the constraints of palaeobiogeographic analyses.

Biogeographical patterns may be expressed in different ways. This chapter will discuss the most commonly used qualitative and quantitative criteria to define biogeographic units, which are widely employed for biogeographical classifications.

4.1 Rationales

In a fundamental analytical review of the main concepts of biogeography, Rosen (1992) has defined 12 rationales according to biogeographic/palaeobiogeographical aims. Six of these rationales are specifically related to palaeobiogeographical aims. Although presented as subjective and probably incomplete, Rosen's analysis provides a theoretical framework which we can use as a guide in palaeobiogeographical analyses. Rationales acknowledge the assumptions which can be made in biogeographical investigations; furthermore, the assumptions of each rationale constrain the choice and quality of data used, as well as the choice of analytical methods. The six Rosen's palaeobiogeographical rationales are summarized below (numbering corresponding to Rosen's original numbering of biogeographical rationales).

4.1.1 Rationale 1

palaeobiogeographical raw data → palaeobiogeographical patterns

Raw data correspond to taxonomic identifications, presence/absence data of particular taxa or ecological categories of taxa at OPGUs specified by their locations in space (geographical occurrence) and time (the age of the deposit in which the fossil occurs). This is the most objective starting point in palaeobiogeographic analysis. This rationale primarily covers descriptive aims through extraction of distributional patterns, definition of areas of endemism, description of ranges of taxa, patterns of provinciality and patterns of taxonomic relationship in relation to areas. Rationale 1 also allows construction of curves of turnover rates or survivorship curves with respect to palaeogeography (e.g. 'hollow curve' patterns) and derivation of geoecological patterns (trophic categories, palaeocommunities, palaeoenvironments, etc.).

4.1.2 Rationale 2

palaeobiogeographical raw data → palaeobiogeographical patterns → biogeographical process

This rationale is aimed at explanation of biogeographical patterns and investigation of biogeographical processes, i.e. the causes of the palaeobiogeographical distributions. Causes may be ecophysiological, geological or evolutionary.

Rationales 1 and 2 represent the 'core' rationales of palaeobiogeography and cover phenetic and causal aims. However, palaeobiogeographical distributions of fossils have also been studied as a means of inferring earth's geography in the past and geological events (Rosen's TECO events). Applied aims need a further step corresponding to the following rationale 3.

4.1.3 Rationale 3

palaeobiogeographical raw data → palaeobiogeographical patterns → biogeographical process → relevant geological and related process

This rationale is aimed at the use of palaeobiogeographical patterns to infer geological history and palaeogeography. Geological interpretations derived from distributional patterns may be different and open to uncertainty because the precision of the geological inference depends on knowledge of relevant biological processes. The fourth step of rationale 3 emphasizes the relationship between geology and biology, but biological processes in biogeography are still so poorly understood that it is necessary to make assumptions about them, and therefore modify rationale 3, in order to make geological inferences.

4.1.4 Rationale 4

rationale 1 + ASSUMPTION OF BIOGEOGRAPHICAL PROCESS → relevant geological and related process

This corresponds to the most common investigative pathway followed in applied marine palaeobiogeography. The biogeographical processes are largely assumed here and not investigated as in rationale 3. Geological history can be inferred using palaeobiogeographical patterns of different kinds, such as endemism, disjunctions, biogeographical assemblages, curves of extinction, origination, survivorship and evolutionary rates, phylogenetic patterns, and palaeoecological groupings of taxa. However, the risk of selecting *ad hoc* assumptions of biological process and making the biogeographical results fit with independent geophysical reconstructions should be avoided. This is extremely important because palaeobiogeography assumes, or contributes to, reconstructions of ancient geographies and TECO events. There is a conflict between rationale 4, which assumes the biogeographical processes, and rationales 2 and 3, which aim to discover them. In other words, the geological processes inferred by rationale 4 cannot be used to infer biogeographical processes.

4.1.5 Rationale 5

rationale 1 + ASSUMPTION OF GEOLOGICAL PROCESS → biogeographical process

Similarly to rationale 2, the aim here is biogeographical because assumed geological events and distributional patterns are used to infer biogeographical processes: investigation of factors of provinciality and its role in diversity regulation, explanations of latitudinal diversity gradients, etc. Although the choice of the assumed geological process may be constrained, in some cases, by the quality of the geological database, assumptions of *ad hoc* processes may be unavoidable because of the poor knowledge of the relationship between TECO and biological processes. This rationale is incompatible with rationales 3 and 4, which infer geological processes, i.e. the same data set cannot be used both to infer biogeographical processes and to make palaeogeographical reconstructions that served for these inferences.

4.1.6 Rationale 6

rationale 1 + ASSUMPTION OF GEOLOGICAL PROCESS → biogeographical process → evolutionary biological or ecological biological processes

This rationale is aimed at investigation of palaeobiogeographical factors in evolutionary or evolutionary–ecological biology, such as the role of provinciality in diversity regulation through time, turnover rates and survivorship patterns in geological or ecological time with respect to palaeogeography. There is incompatibility with rationales 3 and 4.

4.1.7 Risks and difficulties

Rationales 4, 5 and 6 contain assumptions which can generate many of the problems of biogeography, whether or not these are clearly acknowledged. Geological conclusions should be kept separated from biogeographically derived conclusions in order to avoid circularity. The reconstructions of the palaeogeographic positions of the main continental masses in the Lower Palaeozoic (McKerrow and Scotese, 1990) provide a significant example as they largely depend on palaeobiogeographical data. The same fossil data should not be used to infer palaeobiological processes (rationales 5 and 6) as well as to assist in making palaeogeographic reconstructions (rationales 3 and 4) on which these inferences are based. Poor palaeomagnetic constraints do not permit an independent control. Descriptive aims are represented by rationale 1, causal (ecological and historical) aims by rationales 2 and 5, and applied aims by rationales 3, 4 and 6.

4.2 Limits and constraints of palaeobiogeography

The limits of palaeobiogeography depend on aims. However, the incomplete nature of the fossil record constrains palaeobiogeographical analysis, defining its limits at the same time. Although some of these limits appeared in the discussion about the differences between neobiogeography and palaeobiogeography, a detailed discussion on constraints of palaeobiogeographic analysis seems necessary now. Eight kinds of constraints should be considered in palaeobiogeography (Rosen, 1992):

1 taxonomic precision
2 sample size
3 incompleteness of the fossil record and sampling problems
4 stratigraphical correlation and resolution
5 the hypothetical nature of palaeogeographical reconstructions
6 geotectonic setting
7 discontinuity of sample horizons and problems of generalization
8 ecology and taphonomy of fossils.

4.2.1 Taxonomic precision

Sound systematics is necessary to palaeontological studies of all kinds, from biostratigraphy to palaeobiogeography. Good taxonomy is crucial to most studies, particularly those based on historical methods (e.g. cladistic biogeography). Different systematic philosophies, or simply different practices, have a considerable impact on the results of the analysis. Systematics always contains some subjectivity, therefore palaeobiogeographic analyses using fossil data studied by different taxonomists may be more or less biased. A congruent (rather than 'exact') taxonomy is therefore the sole realistic solution which can be obtained after a taxonomic re-evaluation of the whole fossil database by a single author (or by authors of a same 'school'), although this practice may be tedious or even impossible to realize in most cases. A detailed taxonomy is not always necessary for palaeogeographic reconstructions using particular facies-constrained assemblages (e.g. reefal communities): for example, the palaeoposition of displaced terranes may be constrained when the boundaries of past climatic belts are recognized, i.e. if it is only the facies, not the taxa, that matter.

4.2.2 Sample size

Biogeographical patterns are more reliable as numbers of identifiable specimens at any single sample point increase. Differences in sample size can generate false biogeographical patterns, as demonstrated by Koch (1987). The species-frequency distribution fits the log series distribution, thus allowing sample size to be predicted. Koch suggests that infrequent species should be eliminated from biogeographic considerations when species occurrence frequency distribution cannot be calculated. However, other non-statistical similarity methods take into account rare taxa as the main indicators of endemism.

4.2.3 Incompleteness of the fossil record and sampling problems

The incomplete nature of the fossil record has provoked sceptical views over its use in biogeography, even though the representativeness of the fossil database has been demonstrated (Bambach, 1977; Rosenzweig, 1977; Foote and Sepkoski, 1999). However, incompleteness does not prevent the use of fossils in biogeography because they provide a considerable source of information which includes biogeographic signals (Grande, 1985; Rosen, 1992; Hallam, 1994). The latter are certainly incomplete, and probably insufficient for studies based on one particular analytical method, but we must be able to find the best methods to derive the most objective and reliable patterns. Palaeobiogeographical hypotheses would be more rigorous and testable by other workers if based on palaeontological monographs instead of sparse information or provisional faunal lists not followed by the full palaeontological study. Therefore, it is important for the raw data to be validated by published specimens and details of the institution where the material is housed. An incomplete documentation could be produced by insufficient **sampling** in particular areas or in particular stratigraphic intervals. Sampling problems, which should not be confused with problems of sample size, may produce patterns derived from insufficient raw data. For example, the studies of Budd *et al.* (1992, 1994) and Wilson and Rosen (1998) on Caribbean and Indo-West Pacific reef coral faunas respectively have demonstrated that palaeontological sampling was poor or lacking in Palaeogene sediments of both areas. New sampling at close intervals within a stratigraphic framework provided a more accurate database which, in contrast with previous theories (see section 3.5.1) based on an insufficient sampling of the Palaeogene faunas, supports the hypothesis of a relatively recent origin of the coral faunas of these regions.

4.2.4 Stratigraphical correlation and resolution

Neobiogeographic studies are based on synchronous distributional data, corresponding to the present-day time horizon. Palaeobiogeography cannot work with synchronous and 'instantaneous' fossil data: even if the different assemblages from a set of sample localities may well belong to the same bio-chronostratigraphic unit, the resolution of the different rock-dating methods (biostratigraphy, ciclostratigraphy, etc.) can only give us the finest time interval. This is obviously far longer than the rate of the different biological and ecological processes (maintenance) and cannot be compared with the 'instant' view of living distributions (Bennett, 1990). According to the geological period investigated, different temporal resolutions can be obtained; for example, the finest resolution is obtained in the Quaternary. Stratigraphic correlation of fossil assemblages from geographically separated areas must be as precise as possible. Considerable time inconsistencies may result from uncertainties in correlation, thus making the results of the analysis less reliable. On the other hand, the use of different chronostratigraphic scales may also bias correlations, for example the use of different stage names in different parts of the world (e.g. the use of three different stage names, Portlandian, Tithonian, Volgian, for the last stage of the Jurassic period) that do not correlate precisely with one another.

4.2.5 The hypothetical nature of palaeogeographical reconstructions

As already discussed for differences between neobiogeography and palaeobiogeography, palaeogeographic reconstructions are mostly hypothetical. The Earth's surface at a given age is obviously inferred from geological data which cannot assemble contemporaneous palaeogeographic scenarios in synchronous palaeogeographical reconstructions. Even in the cases of the finest chronologic resolutions, palaeogeographic reconstructions refer to given stratigraphical intervals, whose duration is variable according to the different periods of the Phanerozoic and to the rock-dating method employed. It is impossible to plot fossil distributions on palaeogeographical maps with the same confidence that neobiogeographers plot their data on maps of the modern Earth. However, distributional data are commonly plotted on current palaeogeographical maps whose hypothetical basis strongly biases the results of palaeobiogeographical analyses which depend on prior palaeogeographical reconstructions (see Rosen's rationales 5 and 6 above). The use of such plots may entail circularity when our study is aimed at solving palaeogeographic problems from fossil distributions or testing such derived reconstructions (see rationales 3 and 4 above). On the other hand, current palaeogeographical reconstructions cannot be ignored when the aim is to explore the implications of palaeobiogeographical patterns for evolutionary theory. Rosen (1992) addressed these problems and proposed four guidelines to avoid circularity, while maximizing the benefit of palaeogeographic maps leaving open the possibility of their continual revision.

1 A concrete topographical frame of reference with current place-names must be specified for OPGUs, together with the most objective structural and stratigraphic details.
2 For studies aimed at investigation of palaeobiological–evolutionary factors (rationales 5 and 6), details of palaeogeography and geological history should be explicitly specified as assumptions and kept distinct from the objective locality information based on present geography.
3 For the same rationales 5 and 6, relevant uncertainties of a palaeogeographic reconstruction should be indicated together with alternative reconstructions. This is particularly needed, as mentioned above, for Lower Palaeozoic palaeogeographic reconstructions.
4 The use of information based on palaeogeographical maps for palaeogeographical purposes is valid in order to make comparisons between patterns derived from biogeographical data with independently derived palaeogeography or geological history. However, there are no procedures for objectively determining compatibility between independently derived palaeogeographies, and usually subjective evaluations of closeness of match are expressed.

4.2.6 Geotectonic setting

Depending on geological processes that may have affected the palaeomagnetic signal or the structural integrity of particular terranes, some palaeogeographic reconstructions are better constrained than others. Madagascar is a classic example, because it is a constant recognizable physical unit since the Late Jurassic (Dercourt *et al.*, 1993), although its position relative to Africa, Asia and India has changed continuously. On the other hand, a particular region may consist of different tectonostratigraphic units, or terranes, as shown by South and South-East Asia (which consists of accreted Gondwana terranes), Peri-Adriatic alpine chains or the Pacific coast of both North and South America. Cratonic settings most commonly display integrity (absence of thrusting, unfolded and scarcely faulted successions) and provide the best opportunities in studies aimed at inference of palaeobiological and evolutionary processes (rationales 5 and 6).

4.2.7 Discontinuity of sample horizons and problems of generalization

We have already stressed the temporal and taxonomic incompleteness of the fossil record, which is also laterally discontinuous because of geographical gaps. Palaeobiogeographical sampling is limited to samples of the original biotas constrained by taxonomic, preservational and chronostratigraphic losses of information. Therefore, palaeobiogeographers address problems of incomplete sampling and gaps of the fossil records by generalizing areal distributions derived from available data sets of contemporaneous (as far as geological time at different scales is concerned) sample localities. However, criteria for generalization are different, they may be arbitrary or subjective, and may not always be clearly stated. Among generalized areas there are formal biogeographic units (province, realm, etc.), which may provide the basis for generalization, although they are sometimes treated as if they were real sample areas. On the other hand, problems arise when the same name is used for a generalized unit whose definition is based on different criteria. For example, the name 'Tethyan Realm' is currently employed to mean a physically defined marine region, or a climatic–oceanographic latitudinal (tropical) belt, or a belt of high taxonomic diversity, or by its distinctive faunas. Choice of organisms for generalizing areas is subjective and leads to an arbitrary selection of groups of organisms considered biogeographically informative or uninformative. The choice of taxa for defining areas should take into account groups showing the greatest consistency of pattern with each other and the finest geographical resolution. As regards tectonic constraints, generalization is a reasonable working approach when terrane integrity can be used to generalize records to the terrane as a whole (Smith, 1988). In the case of collages of terranes, each terrane can be treated as an OPGU in order to investigate biogeographical history between component terranes (see Chapter 5 for a summary of Smith's work on east-Asian terranes).

4.2.8 Ecology and taphonomy of fossils

Syndepositional or post-depositional factors may alter the palaeobiogeographical signal through resedimentation, reworking, etc. Therefore different ecological groups should be identified, provided that the ecology of the organism is known. This problem is dramatic in extinct groups, whose ecology is hypothetical. The taphonomic processes could concentrate fossils from different stratigraphic intervals to that of the formation of the bed containing the assemblage under study (taphonomic condensation *sensu* Fernandez-Lopez, 1991). This kind of faunal concentration can also produce an anomalous association of organisms belonging to different ecological assemblages.

4.3 Definition of biogeographic units: qualitative criteria

The end-points of spatial ranges of extant taxa tend to cluster in coincidence with geographical discontinuities (oceans, mountain chains, climatic belts, etc.), thus definining boundaries of areas of endemism. Clusters of range end-points of fossil taxa should define, similarly, palaeogeographical discontinuities and boundaries of past areas of endemism. According to Jablonski *et al.* (1985), the basic biogeographic unit should be the range of a single taxon (i.e. the geographical area it naturally occupies at any one time). However, more inclusive units which embody assemblages of taxa and their ranges have always been defined by biogeographers who, historically, classify and describe distributions through the use of biogeographic units. Different techniques are used for biogeographical classification. Quantitative techniques are more used in neobiogeography than in palaeobiogeography. Although in the palaeobiogeographic literature numerous attempts to use similarity cofficients exist, the incomplete nature of the fossil record has been criticized and more complex statistical elaborations have been devised. Areas of endemism, or biogeographic units, are mostly defined in palaeobiogeography on qualitative grounds.

The concept of biotic province is a long-established one since the times of Sclater, de Candolle and Wallace (Nelson and Platnick, 1981). In palaeobiogeography Neumayr (1883) introduced the zoogeographic **province**, defined as a large area characterized by a common particular fauna, caused by geographic position (barriers, latitude, etc.) but independent of facies. Uhlig (1911) introduced a more inclusive rank, the realm, and stressed the independence of facies. Thus, according to their original definitions at least, biogeographic units are based on endemism and should have a historical meaning. However, different definitions and concepts of biogeographic units have been proposed by a number of authors. But there is no doubt that this notion has been extremely important in the history of biogeography because it stressed the importance of endemicity and developed the idea of relations between regions (Rosen and Smith, 1988).

4.3.1 Concepts and definitions of biogeographic units in palaeobiogeography

As discussed in Chapter 2, endemism is a relative concept which implies a hierarchical arrangement of areas of endemism, thus explaining the use by different authors of other related concepts: realm, region, biome, etc. Accordingly, different classifications of biogeographic units of different rank exist. Kauffmann (1973) observed that neobiogeographers mainly use the units (decreasing rank) planetary biota, realm, region, province or subregion, subprovince, endemic centre, community, and that some authors abandoned 'realm' or replaced it with 'region', intended as a synonym. He also noted that some specialists proposed percentages of endemism to define biogeographic provinces. On the basis of a survey of the biogeographic literature available at that time, Kauffmann realized that no guidelines existed for ancient biogeographic units and that Late Cenozoic workers apply the units used by neobiogeographers. Therefore he proposed a system for differentiation of biogeographic units, which he directly applied to Cretaceous bivalves, based on percentages of endemism (all calculated exclusive of cosmopolitan taxa): endemic centres, 5–10% endemism; subprovinces, 10–25%; provinces, 25–50%; regions, 50–75%; and realms, >75% endemism.

Valentine (1973), who defined provinces as regions in which communities maintain characteristic taxonomic compositions, stated that they represent the basic descriptive biogeographical unit and observed that there is no special reason to employ any particular arbitrary level of endemism because a province could possess no endemic species at all and yet have distinctive communities (Fig. 4.1).

McKerrow and Cocks (1986) proposed the use of the term 'province' for regions separated by physical barriers to migration and 'realm' for climatically controlled faunas and floras, thus dismissing the hierarchy commonly employed by most authors. This proposal, which merely introduces a different hierarchy (climate acts at a larger geographical scale than a barrier), has not been followed much.

Confusion about the principles of classification and terminology still exists. Therefore Gerd Westermann has recently founded the 'Friends of Palaeobiogeography', an international group of specialists in different phyla throughout the Phanerozoic, with the primary topic of marine palaeobiogeographic classification (Westermann, 2000a). A sample of palaeobiogeographers has expressed a consensus on principles of classification and nomenclature of biogeographic units or biochores ('group of similar biotopes, the largest division of animal and plant environments' according to the *Oxford English Dictionary*). The basic guidelines, or recommendations, are as follows.

1 *Ranks (or tiers) of biogeographic units*: realm, subrealm (optional), province, subprovince (optional). The term 'region' should be used informally rather than as a substitute for subrealm. Russian authors

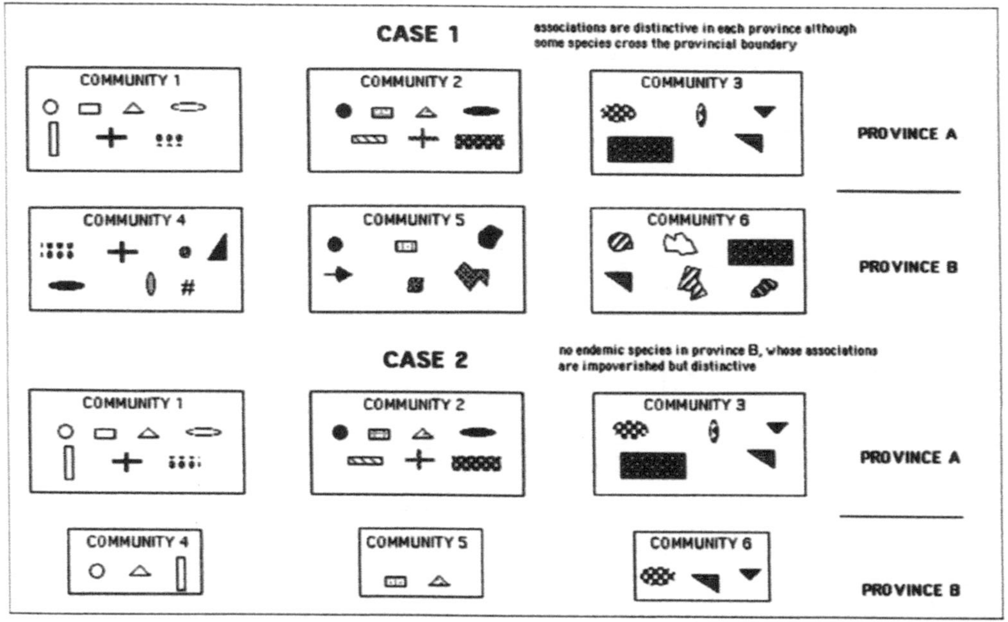

Figure 4.1 Two patterns of provinciality (objects represent species). (Case 1) Communities 1, 2 and 3 of 'Province A' mainly contain different species from communities 4, 5 and 6 of 'Province B': although some species range across the provincial boundary, the associations are distinctive in each province. (Case 2) Theoretically possible pattern characterized by the lack of endemic species in 'Province B', which displays an impoverished, but distinctive, association compared to 'Province A'.

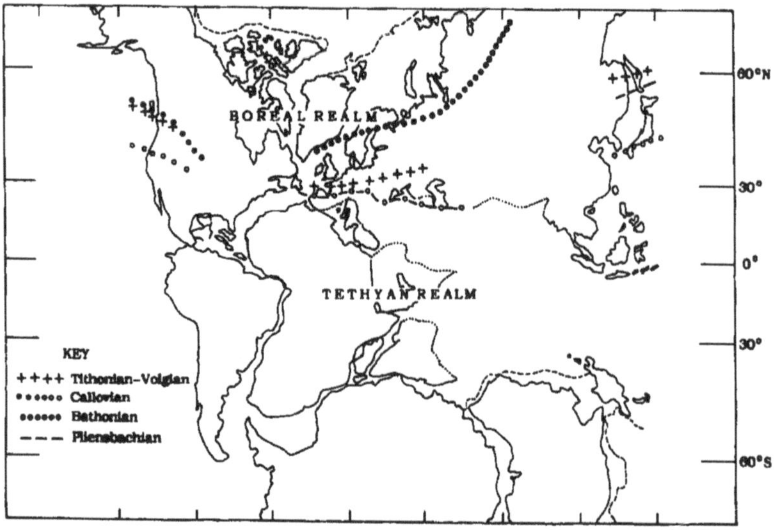

Figure 4.2 The approximate boundaries of the Tethyan and Boreal Realms at different times in the Jurassic.

Source: Hallam (1975). Reproduced with permission of Cambridge University Press.

use an additional rank, variably named kingdom, superrealm or belt, at the top of the classification.

2 *Definition of biochores*: they should ideally be based on biota, not on single taxa.
3 *Biochores as dynamic units*: they change in range (areal extent) and rank through time; some may disappear at mass-extinction and major geotectonic or eustatic events. The shifting of the boundary between the Boreal and Tethyan Realms in the Jurassic provides an example of change of biochore (Fig. 4.2).
4 *Biochore nomenclature*: names are geographic/geologic, not taxonomic. Priority 'rules' of synonymy and homonymy should be applied to biochore names, not rigorously but with common sense.

The question arises of how to define biochores and their boundaries. Provinces and biochores of all ranks have often been defined qualitatively, averaging similarity or difference (endemicity) of faunas.

This introduces subjectivity and different authors may have different definitions of the same biotic unit. For example, there is strong disagreement between authors as to whether a bipolar distribution of high-latitude taxa since the Jurassic existed or not to justify the introduction of an Austral Realm (whose distinct faunas should be extracted from the Tethyan Realm) as opposed to the Boreal Realm (an exhaustive discussion of this problem can be found in Westermann (2000b)). On the other hand, quantitative definitions through statistical and multivariate treatments of raw biogeographic data and/or index of similarity ('phenetics') have been used. These more rigorous techniques are strictly process-neutral and certainly provide less subjective identifications of biogeographic units, though they do not resolve either problems of objective scaling of biochore ranks or diversity-based differences between adjacent biochores with all taxa of the less diverse biochore common to both.

Biochore boundaries are placed according to the range limits of constituent taxa at any particular time. However, geographical ranges overlap especially with the number and taxonomic rank of taxa considered. Furthermore, ranges change through time. Boundaries have been defined subjectively as well as statistically by clusters of end-points of ranges of individual taxa. Practical difficulties exist for recognition of range boundaries of species distributions. In fact the distinction of true resident populations from parademic and non-breeding and not recruited 'pseudopopulations' (see Chapter 2) is usually difficult and should be based on detailed palaeobiological studies showing the presence *in situ* of juvenile individuals and representatives of both sexes.

Boundary problems may also be related to the choice of taxa to define biochores. According to guideline 2 above, biochores should ideally be based on biota, not single taxa. However, use of multitaxial biochores may produce incongruencies between the range-limits for different higher taxa. In fact, range differences between higher taxa of very different ecological characteristics, for example benthic and pelagic organisms, can be really unreconciliable. Pelagic organisms have much larger distributional areas than benthic ones and their biochores will be geographically wider and numerically reduced. The present-day distribution of planktonic organisms roughly coincides with latitudinal belts. However, factors other than climatic ones may locally control the distributions, as evidenced by the high radiolarian concentrations in zones of important nutrient input from continents. Box 4.1 gives an example of an attempt to resolve some of the above issues.

Box 4.1 A case study: the Tithonian and its ammonite faunas

I have selected from my personal research activity an example of conflicting biochore definition to show both problems of biochore scaling and how a biochore can be defined on the basis of different starting points. The example concerns the ammonite palaeobiogeographic distribution in the Mediterranean and European areas during the Tithonian (Cecca, 1999). First, it is interesting to emphasize that the Tithonian is a special time-horizon for palaeobiogeography. In fact endemism was extremely strong, making biochronologic correlations so difficult (and even impossible) that three stage names, Tithonian, Portlandian and Volgian, each one being characterized by a typical fauna, are used for the last stage of the Jurassic. In particular the Tithonian is used in the Tethyan and Indo-Pacific areas, the Portlandian in southern England and in the Paris Basin, the Volgian in the Russian platform and in Siberia.

As illustrated in Chapter 3, two realms, Boreal and Tethyan, are identifiable in Europe. Their boundaries fluctuated during the Jurassic. In the Tethyan Realm different faunas have been used to identify provinces. The Mediterranean Tethys was a branch of the Tethys ocean (Dercourt *et al.*, 1993) and in Fig. 4.3 (see also Fig. 2.4) it corresponds to the complex palaeogeographic domain between Europe and Africa. On the other hand, the term 'Mediterranean Province' corresponds to a palaeobiogeographic unit which has been used with different meanings according to time-horizon and fossil group considered. Concerning the Tithonian and taking into account ammonite faunas, differences in diversity allow faunal distinctions immediately. In southern Spain, Italy, northern Africa, and north-western Hungary, representatives of eight families occur whereas in southern France, southern Germany, Romania and Caucasus this number decreases, fluctuating from six to one in the different time-horizons (for example ammonite biozones) of the Tithonian. The more diverse fauna defines a palaeobiogeographic entity that we call the Mediterranean Province, whereas the less diverse one identifies a second entity called the Sub-Mediterranean Province. This is the classification originally proposed in a seminal paper by Enay (1972), who mainly based these provinces on faunal differences (degree of endemism) more than diversity. The diversity-based distinction is mine (Cecca, 1986, 1992). Note that subsequently Enay (1980b) proposed a unique Mediterranean Province subdivided in the Mediterranean s. str. and Sub-Mediterranean Subprovinces. It is also worth noting that both Enay and I have worked subjectively, averaging faunal differences and/or similarities. This illustrates how two authors working on the same faunas and having a similar background (I was one of Enay's students)

Box 4.1 continued

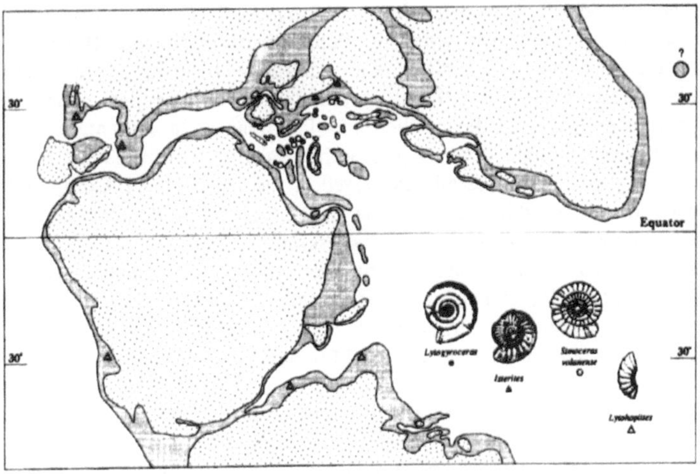

Figure 4.3 Palaeobiogeographic distribution of selected taxa at the end of the Early Tithonian. Note the endemic distribution of *Lytogyroceras* and *Simoceras volanense* in the Mediterranean areas and of *Isterites* in the Sub-Mediterranean ones.

Source: Cecca (1999). Reproduced with permission of Elsevier Science.

may use different criteria to define the same biogeographical entity (keeping apart the different ranking). But why did I develop this diversity-based distinction?

On the basis of a detailed revision of Mediterranean Tithonian ammonites I first realized that, for a given time-horizon such as an ammonite zone, the distinction between Mediterranean and Sub-Mediterranean faunas (not provinces!) was primarily ecological: the former lived in distal, pelagic habitats connected with rifted continental margins while the latter essentially lived in epicontinental platforms and intracratonic basins. Palaeotectonic causes are not (or not entirely) responsible for the existence of these two faunas: in fact, if the Sub-Mediterranean fauna is recorded only on the northern margin, the Mediterranean fauna existed on both the southern and northern (in southern Spain) margins of the Mediterranean Tethys. On the other hand, long-established environmental differences led to historical isolation of distinct faunas and, at the scale of the Jurassic period as a whole, this faunal distinction

Box 4.1 continued

often coincides with evident phases of provincialism which correspond to vicariant events. Yet, when the faunal distributions are studied at shorter chronologic intervals (e.g. ammonite biozones), provincial subdivisions may have different meanings or even disappear. Then, I secondly realized that during the Early Tithonian all families and virtually all genera might occur within both faunas, endemism having been extremely rare and generally expressed at the specific level. Compared to the diverse Mediterranean fauna s. str., the Sub-Mediterranean one appears an impoverished assemblage sometimes enriched by sporadic occurrences of species that were best represented in the distal, pelagic habitats of the Mediterranean areas. I concluded that, at this time-scale, the two faunas mainly correspond to ecologically controlled communities (biomes *sensu* Dommergues and Marchand, 1988) more than assemblages separated by vicariant processes.

Here, we should note that a different level of temporal perception allows the observation of different causal factors: vicariance can be invoked at the scale of the Jurassic time interval whereas ecological control is mainly invoked working at shorter time-intervals such as the substage or the ammonite biozone. This does not mean that vicariant processes could not take place at these shorter time intervals: if the analysis is carried out at specific level we certainly could distinguish endemic species, particularly in the taxonomically puzzling family Ataxioceratidae. Furthermore, I recognized during the transition between Early and Late Tithonian a clear phase of faunal divergence with related endemism of generic level. The problem of the distinction of a Sub-Mediterranean palaeobiogeographical entity of province rank for the time interval spanning from the last to the first ammonite zones of the Early and Late Tithonian respectively has not been addressed, though it could underline a historic phenomenon of biogeographic divergence.

In summary, conflicting definitions of these palaeobiogeographic units are shown by this example because both diversity-based and endemism-based criteria have been used. Yet distinct processes seem to have acted at different times within the Tithonian. The phase of faunal divergence at the Early/Late Tithonian transition can be tested by the independent evidence of an important marine regression (Haq *et al.*, 1988; Hallam, 1988; Fourcade *et al.*, 1991) resulting in severe contraction of ammonite habitats.

4.3.2 Biota, biome and their overlaps

The term **biota** indicates the whole association of organisms that live in the same area. Floras and faunas share a common historical succession of evolutionary events which have taken place in this area. Therefore the term 'biota' is reserved to communities having a historical significance. The term **biome** was introduced by botanists for vegetal assemblages corresponding to well-defined conditions and is essentially used by ecologists. It has been popularized among palaeontologists by Sylvester-Bradley (1971): for example in the marine realm coral-reefs, level-bottom and planktonic biomes, or many others correspond to biomes. Biomes certainly have a geographic component which is exclusively controlled by ecological factors. Biomes have been considered biogeographic units corresponding to communities with similar functional adaptations by Dommergues and Marchand (1988, p. 354). These authors applied the concepts of biota and biome to the repeated episodes of range expansions of Boreal and Tethyan ammonites during the Jurassic, in order to distinguish ecological and historical components.

According to Hallam (1994, p. 22), the classification of biomes is irrelevant to biogeography, which is concerned with the distribution across the globe of similar types of biome, or the geographic restriction of some biomes. However, the distinction of these two concepts is important for applications of palaeobiogeography to palaeogeographic reconstructions. The exotic character of fauna and/or floras of suspected terranes (Chapter 8) must be based on historical biogeographic differences with the biotic component of the mainland. In fact, an analysis aimed at inference of the original palaeogeographic position of a suspected terrane may prove misleading or inconclusive if based on biomes.

Dommergues and Marchand (1988) have also introduced the terms **interface** and **intergradation** (see Fig. 3.4). The former indicates the overlap of species from two biotas, whereas the latter corresponds to the overlap of two biomes. However, the practical distinction of interface and intergradations is not clear. When barriers fall and species of different biogeographic origins merge, the areas of overlapping usually show a mix of communities with different historical origins and adaptation to similar ecological conditions, i.e. what Dommergues and Marchand (1988, p. 360) called 'interface/intergradation complexe'. The terms **biotic interchange** (Vermeij, 1991) and **ecotone** also indicate the overlap of faunas and/or floras of different biogeographic origins and are equivalent to 'interface/intergradation complexe'.

4.4 Quantitative criteria ('phenetics')

4.4.1 Binary similarity coefficients

Faunas at sample localities may be compared through the use of mathematical similarity coefficients ('phenetics'). The calculation of the values of these coefficients generally starts from presence or absence (binary method) of taxa in the faunal lists. The values are compared with each other by constructing a matrix which can be treated further by multivariate methods such as cluster analysis and ordination. All these methods provide a quantitative description of faunal similarity and are totally process-neutral, though the results are expected to express somehow the effects of processes.

Various similarity coefficients are available (Cheetham and Hazel, 1969; Hazel, 1970; Shi, 1993). As we will see in the next section, the value of 'conventional' coefficients, as opposed to probabilistic similarity coefficients, has been criticized in palaeobiogeography, though recent studies claim they can produce significant results (Shi, 1993). Nevertheless the conventional similarity coefficients have been widely used to compare faunas of different areas and it is necessary to illustrate them here. In the palaeobiogeographical literature the most widely employed binary similarity coefficients are Jaccard, Simpson and Dice coefficients. These coefficients have been considered to perform better than others for biogeographic purposes after tests and comparisons of 39 conventional binary similarity coefficients (Shi, 1993). They have mostly been used to generic level because treatment to the specific level appears to be totally inadequate for taxonomic reasons (Hallam, 1977). However, Campbell and Valentine (1977) stressed that treatment to the generic level may also be difficult, even in the case of systematic classifications by a group of palaeontologists sharing a common taxonomic tradition. Furthermore, these coefficients have been used to compare faunas of different land masses or (with circularity, I argue) to assess palaeobiogeographic provinces subjectively defined beforehand.

Jaccard coefficient, J, is the ratio of C, the number of common taxa between two areas (or provinces), and the total number of taxa occurring in both areas:

$$J = C/(n_1 + n_2 - C) \qquad (1)$$

where n_1 and n_2 are the numbers of taxa in the smaller (less diverse) and the larger (more diverse) faunas respectively. But n_1 and n_2 may also be used as provinces, regions, land masses, etc.

Simpson coefficient, which was devised to minimize the effect of unequal size of the two faunas being compared, is expressed by the ratio of common taxa between two areas (or provinces) with the number of taxa present in the smaller (less diverse) fauna:

$$S = C/n_1 \qquad (2)$$

Areas (or provinces) can be considered 'perfectly' similar on the basis of Simpson coefficient no matter how different they are in diversity: if all taxa of the smaller fauna also occur within the larger one, values of Simpson coefficient are equal.

According to Valentine (1973), the Simpson coefficient stresses similarity whereas Jaccard's emphasizes differences between faunas because it is more sensitive to differences of diversity. Simpson (1960) observed that if two faunas were approximately equal in size, the Jaccard coefficient would be satisfactory. Fallaw (1979) noted that in the case of a great difference in number of taxa, the larger fauna would distort the resulting value so that the degree of relationship between the two faunas would be obscured. This author supported his statement with an extreme example: 'suppose that two faunas had been identical in life, but, as a result of poor preservation or lack of outcrops in one area, n_1 is 100 and n_2 is 1,000 in the known fossil assemblages. The Jaccard value for this situation would be 0.1, even though every opportunity of having identical taxa in the known faunas was realized. The Simpson coefficient in this case would be 1.0' (Fallaw, 1979, p. 1029). However, Hughes (1973) observed that if one sample is very small the Simpson coefficient may produce unstable results because small increases of information may lead to disproportionately large effects on the result. Therefore this author recommended the use of Jaccard or Dice coefficients instead of Simpson's, and that greater importance be placed on well-known large faunas than on poorly known small faunas. Archer and Maples (1987) recommended that the Jaccard coefficient should be used with caution. More recently Shi (1993) has strongly recommended the Jaccard coefficient because it meets nearly all the conditions of his comparative test of binary coefficients; furthermore, its use would strongly reduce distortions of information prior to the treatment of the raw data by means of multivariate methods such as cluster analysis.

The Dice coefficient, similarly to Jaccard's, is suitable if the two faunas are of equal size:

$$D = 2C/n_1 + n_2 \qquad (3)$$

According to Shi (1993), the Jaccard and Dice coefficients are expected to produce the most meaningful results.

Johnson's provinciality index PI is also a widely applied binary coefficient because it provides a method to compare regions with samples of differing sizes:

$$PI = C/2E \tag{4}$$

where C is again the number of taxa in common between two regions and E is the number of taxa endemic to the region of lower diversity. It is worth noting that sample size is not taken into account in this equation.

I have selected two examples of the application of Jaccard and Simpson coefficients in palaeobiogeographical analyses.

4.4.1.1 Faunal similarity and the widening of the Atlantic Ocean

The opening of the Atlantic Ocean, deduced by geological and geophysical evidence, resulted in the creation of endemic eastern and western Atlantic faunas through biogeographical separation of a former biota. This leads to the prediction that this vicariant pattern was achieved through a progressive decrease of faunal similarity between different faunal groups on opposite sides of the Atlantic since the Early Mesozoic (Hallam, 1973; Kauffman, 1973). Coates (1973) has clearly demonstrated the progressive divergence amongst tropical neritic faunas, corals and bivalves including rudists, of the Tethyan Realm, with the appearance of a distinct Caribbean province. This is clearly demonstrated also by the progressive, slow expansion of large benthic foraminifers from the Mediterranean Tethys to the Caribbean with the appearance of endemic genera since the Albian and, more definitely, from the Campanian (Fourcade and Michaud, 1987). Fallaw (1979) analysed the occurrence of marine invertebrate genera in central and eastern North America and in western Europe and northern Africa by means of binary similarity coefficients. Values of Simpson and Jaccard coefficients, derived from lists of genera in eastern and western areas, were given for different time intervals from the Early Triassic (i.e. prior to the first rifting phases) to the Neogene. In accord with the progressive widening of the Atlantic Ocean, decreasing commonality during the Mesozoic and Cenozoic is clearly shown by values of Simpson coefficient, while Jaccard coefficient contradicts the predicted decrease of similarity. The author concluded that among the different binary coefficients Simpson's one appears to be much more effective in

Table. 4.1 Data on Mesozoic trans-North Atlantic commonality for pelecypods, with Jaccard and Simpson coefficients (from Fallaw, 1979).

Time interval	*Genera in western area*	*Genera in eastern area*	*Common to both*	*Jaccard coefficient*	*Simpson coefficient*
Late Cretaceous	167	188	92	0.35	0.55
Early Cretaceous	78	130	59	0.40	0.76
Late Jurassic	19	101	17	0.17	0.89
Middle Jurassic	18	96	16	0.16	0.89
Early Jurassic	13	73	12	0.16	0.92
Late Triassic	17	78	13	0.16	0.76
Middle Triassic	7	40	6	0.15	0.86
Early Triassic	6	24	5	0.20	0.83

measuring the degree of faunal similarity when tested against faunas having a large discrepancy in size and size changes through time. An example may help us to understand this conclusion (Table 4.1). One can ask why Early Triassic faunas should be considered as being more similar than the Late Cretaceous ones since there are only five genera in common in the former epoch and 92 in common in the latter. Fallaw's answer is that out of six chances of having the same genera in the Early Triassic, five are successful, but only 92 out of 167 are successful in the Late Cretaceous faunas. Therefore the Simpson coefficient does not seem to be overwhelmed by faunal size differences and changes in size through time.

The effectiveness of the Simpson coefficient has been claimed once more by Fallaw and Dromgoole (1980) in a study of faunal similarity across the South Atlantic among Mesozoic and Cenozoic invertebrates. In this study the authors have correlated values of Simpson coefficient versus the ocean basin average width of the South Atlantic (Fig. 4.4) and determined the coefficient –0.96 of the linear correlation for the Simpson coefficient (SC) versus ocean basin width, significant at the 0.001 level for a one-tailed test. On the other hand, the linear regression analysis of the data yields the equation:

$$SC = 0.42 - 6 \times 10^{-5} \times \text{width (km)} \quad (5)$$

The strong negative correlation between trans-Atlantic faunal similarity and the average width of the ocean basin led Fallaw and Dromgoole to suggest that sea-floor spreading had a strong effect on invertebrate dispersion between land masses on opposite sides of the ocean. However, this conclusion cannot apply to the whole invertebrate fauna

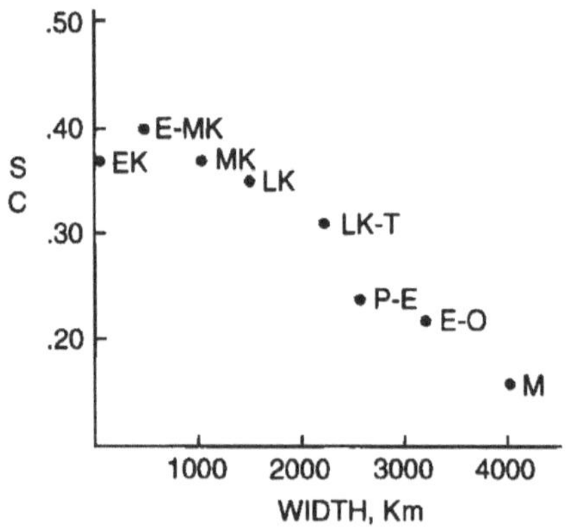

Figure 4.4 Values of the Simpson coefficient (SC) plotted against average width of the South Atlantic ocean basin. M = Miocene, E-O = Eocene–Oligocene, P-E = Palaeocene–Eocene, LK-T = Late Cretaceous, MK = Middle Cretaceous, E-MK = Early to 'Mid' Cretaceous, E-K = Early Cretaceous.

Source: Fallaw and Dromgoole (1980). Reproduced with permission of University of Chicago Press.

because species characterized by long larval stages are found on both sides of the modern Atlantic ocean (Scheltema, 1979).

4.4.1.2 Faunal similarity and East Pacific terranes

Aberhan (1998) analysed the similarity between Early Jurassic pectinoid bivalve faunas of western Canadian allochthonous terranes (Wrangellia and Stikinia) and autochthonous boreal and austral intracratonic basins using both Dice and Jaccard coefficients and cluster analysis. The values of the Dice coefficient are shown in Fig. 4.5. The palaeotectonic reconstruction was based on numerous lines of evidence and the hypotheses about the palaeogeographic position of these terranes are discussed further in Chapter 8. As far as faunal similarity patterns are concerned, Aberhan (1998) proposed three complementary interpretations assuming that similarity coefficients correctly mirror geographical distances.

1 The distance between the terranes Stikinia and Wrangellia was relatively small and during the Hettangian–Pliensbachian interval the former was closer to the continent than the latter. In the Toarcian–Early Aalenian high similarity between Wrangellia and

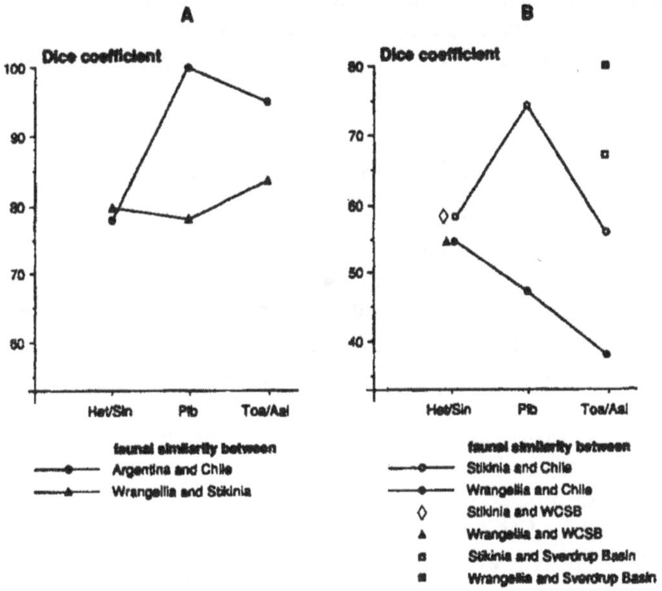

Figure 4.5 Curves showing similarities of Early Jurassic pectinoids of regions of the Eastern Pacific margins as expressed by the Dice coefficient. In (A), high similarities between Argentina and Chile and between Wrangellia and Stikinia suggest the close geographic proximity of these terranes. In (B) patterns of similarities suggest that Stikinia was closer to the craton that Wrangellia as well as its northward displacement after the Pliensbachian. Strong faunal affinities of Wrangellia to the cratonic Sverdrup Basin (Arctic Islands) in the Toarcian–Early Aalenian are indicative of a relatively high, northern palaeolatitude at that time. Abbreviations: Het, Hettangian; Sin, Sinemurian; Plb, Pliensbachian; Toa, Toarcian; Aal, Aalenian; WCSB, Western Canada Sedimentary Basin.

Source: Aberhan, 1998. Reproduced with permission of the Society for Sedimentary Geology.

the cratonic Sverdrup Basin (Arctic Islands) is suggestive of close geographic proximity.

2 From Pliensbachian to Toarcian–Early Aalenian, Stikinia moved considerably northward and, at this time, a position in the northern hemisphere appears to be more reliable.

3 On the basis of the sole similarity coefficients the interpretation of the latitudinal position of Wrangellia in the Pliensbachian is unclear, though relative proximity to Stikinia can be inferred. The strong similarity of Wrangellia and the Sverdrup Basin in the

Toarcian–Early Aalenian is indicative of relatively high, northern palaeolatitudes. Therefore, similarly to Stikinia, substantial northward migration of Wrangellia relative to the craton is assumed for the late Early Jurassic.

4.4.1.3 Comparability of modern and ancient marine provinces

Due to difficulties in recognizing objectively the species level in the fossil record, the fundamental taxon in palaeobiogeographic studies is the genus, or even the family in some cases. Campbell and Valentine (1977) addressed the question of whether modern marine provinces, recognized by their species content, are still distinctive or not at these higher taxonomic levels. The Jaccard and Simpson coefficients from bivalve and gastropod faunas of six eastern Pacific contiguous and five spatially separated modern provinces were computed by these authors, who used these coefficients because of their rather extreme properties compared to other similarity coefficients.

Campbell and Valentine recognized that the present-day spatially contiguous provinces (Fig. 4.6) are still distinct at the generic level, and in most cases at the family level, and concluded that palaeoprovinces defined on the basis of distributional patterns of genera and families are clearly comparable to modern provinces based on the species level. The values of the Simpson coefficient are higher at the family than at the genus level, the more marked differences being between Arctic and Panamanian (Tropical) provinces. In such a case many palaeobiogeographers would not distinguish provinces among regions characterized by such high faunal similarities, even considering that Simpson's coefficient de-emphasizes differences. However, faunal similarities are much lower when the Jaccard coefficient is used and the differences between the genus and the family levels are slight. In fact the latter coefficient is sensitive to diversity differences which have less effect on the coefficient as the total number of taxa decreases at higher taxonomic levels. In other words, the values of the Jaccard coefficient are lower for genera than for families because the total number of taxa appears in the denominator of equation (1).

If longitudinally separated provinces are analysed (Fig. 4.7), the Jaccard coefficient suggests that separated provinces are quite distinct at both generic and family levels, while they are certainly less distinct for the Simpson coefficient. Despite latitudinally contiguous provinces having different climatic regimes, longitudinally separated provinces are more distinctive at these taxonomic levels. At the specific level contiguous eastern Pacific provinces share on average nearly half of their

a) Jaccard

	ARCTIC	ALEUTIAN	OREGONIAN	CALIFORNIAN	SURIAN
ALEUTIAN	.41				
OREGONIAN	.24	.55			
CALIFORNIAN	.16	.38	.70		
SURIAN	.12	.28	.50	.68	
PANAMANIAN	.08	.17	.32	.42	.51

a) Simpson

	ARCTIC	ALEUTIAN	OREGONIAN	CALIFORNIAN	SURIAN
ALEUTIAN	.92				
OREGONIAN	.86	.92			
CALIFORNIAN	.71	.82	.91		
SURIAN	.53	.63	.73	.82	
PANAMANIAN	.46	.61	.73	.79	.93

b) Jaccard

	ARCTIC	ALEUTIAN	OREGONIAN	CALIFORNIAN	SURIAN
ALEUTIAN	.58				
OREGONIAN	.36	.72			
CALIFORNIAN	.32	.62	.85		
SURIAN	.30	.60	.76	.85	
PANAMANIAN	.21	.46	.60	.68	.71

b) Simpson

	ARCTIC	ALEUTIAN	OREGONIAN	CALIFORNIAN	SURIAN
ALEUTIAN	1.00				
OREGONIAN	.98	.97			
CALIFORNIAN	.96	.95	.97		
SURIAN	.89	.91	.89	.94	
PANAMANIAN	.82	.91	.91	.92	.97

Figure 4.6 Jaccard and Simpson coefficients for bivalve and gastropod faunas of six eastern Pacific contiguous modern bioprovinces: (a) generic level; (b) familial level.

Source: Campbell and Valentine (1977). Reproduced with permission of the Paleontological Society.

species, whereas all longitudinally separated provinces are overwhelmingly distinctive in species composition.

4.4.2 Cluster analysis

Two broad groups of multivariate methods as applied to palaeobiogeography may be considered: classification and ordination. The former refers to recognition of discrete groups within a data set and therefore assumes discontinuities and cluster analysis is a particularly useful method; the latter assumes continuities and is useful for displaying interrelationships of data points through ordering and arrangement of samples or variables (taxa in our case). The two methods have

a)

Jaccard

	OREGONIAN	CALIFORNIAN	VIRGINIAN	CELTIC
CALIFORNIAN	.70			
VIRGINIAN	.25	.21		
CELTIC	.29	.27	.30	
MAUGEAN	.09	.10	.06	.09

Simpson

	OREGONIAN	CALIFORNIAN	VIRGINIAN	CELTIC
CALIFORNIAN	.91			
VIRGINIAN	.65	.65		
CELTIC	.49	.52	.67	
MAUGEAN	.23	.23	.29	.24

b)

Jaccard

	OREGONIAN	CALIFORNIAN	VIRGINIAN	CELTIC
CALIFORNIAN	.85			
VIRGINIAN	.53	.48		
CELTIC	.60	.61	.49	
MAUGEAN	.57	.63	.47	.54

Simpson

	OREGONIAN	CALIFORNIAN	VIRGINIAN	CELTIC
CALIFORNIAN	.97			
VIRGINIAN	.93	.93		
CELTIC	.79	.85	.83	
MAUGEAN	.73	.74	.88	.75

Figure 4.7 Jaccard and Simpson coefficients for bivalve and gastropod faunas of five separated modern bioprovinces: (a) generic level; (b) familial level. Note that two contiguous bioprovinces (Oregonian and Californian) have been included for comparison.

Source: Campbell and Valentine (1977). Reproduced with Permission of the Paleontological Society.

different characteristics and drawbacks, and their use is more or less appropriate according to the purpose of the study and the nature of the data. Commonly, binary similarity coefficients are used for analysis of multivariate data, especially in conjunction with hierarchical cluster analysis, which is frequently applied in palaeobiogeographic studies as it groups areas with similar biotas hierarchically. Cluster analysis may be recommended for large-scale palaeobiogeographic studies where binary data sets are employed (Shi, 1993).

The hierarchically grouped clusters correspond to quantitatively defined groups of localities which can be matched with biogeographic units.

Multivariate methods start with the construction of symmetrical matrices of similarity coefficients in the form of comparisons, either sample locality by sample locality (intersite) or of similarities between taxa occurring across sites. The former is called Q-mode analysis and the latter R-mode analysis. The Q-mode analysis is employed in palaeobiogeography, whereas the R-mode is useful in palaeoecological studies because it produces clusters of recurrent species (biofacies) that may represent communities (Valentine, 1973). Most palaeobiogeographic studies use a single-linkage clustering technique where the distance between two clusters is taken to be the distance separating the closest pair of points. The so-called unweighted pair-group method using arithmetic averages (UPGMA) and the weighted pair-group method using arithmetic averages (WPGMA) may define the arithmetic average distance between clusters. The results are presented graphically in the form of two-dimensional dendrograms. It is common in palaeobiogeographic quantitative studies to find cluster analysis of the distribution data which uses Jaccard's coefficient and unweighted pair groups in the Q-mode.

The results of cluster analysis should be compared with results obtained from other analytical techniques because all methods of biogeographical analyses have their advantages and disadvantages (Rosen and Smith, 1988). For application to palaeobiogeographic reconstructions, the results can be compared with independent geologic and/or geophysical evidence (Smith, 1988, 1992; Fortey and Mellish, 1992).

4.4.3 Probabilistic methods

The use of binary similarity coefficients in palaeobiogeography has been criticized by Henderson and Heron (1977). Palaeontologists, hence palaeobiogeographers, deal with samples of original populations and study similarity between the samples by consideration of the number of shared taxa. But what is the relation between sample similarity and population similarity? In this respect, Henderson and Heron provided a diagrammatic representation (Fig. 4.8) of the interrelationships between two hypothetical populations, the samples drawn from them and finally the taxa shared between the samples, i.e. what we actually observe in palaeontology.

Fig. 4.8 shows that the number of common taxa, C, cannot be used to assess similarity between the two populations unless values of N are known. In fact fossil samples allow us to measure C and n but the measure of similarity also depends on N, which is unknown. Let V be the number of taxa shared by the original population of Fig. 4.8B. The equation:

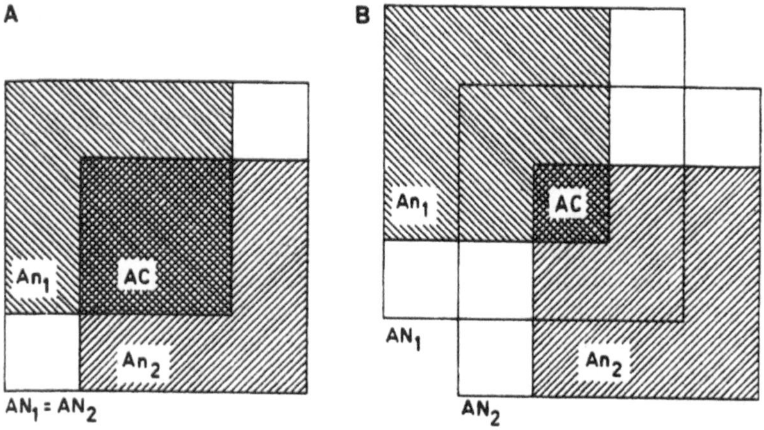

Figure 4.8 Interrelationship of two hypothetical populations, samples drawn from them and shared taxa. Large squares represent original populations of size N_1 and N_2, whose taxonomic constitutions are AN_1 and AN_2; smaller ruled squares are samples of the original populations of sizes n_1 and n_2, whose taxonomic constitutions are An_1 and An_2; the overlap of the small squares represents the taxa shared between An_1 and An_2, whose number is C, and AC is its taxonomic composition. In case A, at left, AN_1 and AN_2 are identical (= identical taxonomic constitution): C varies for different values of n_1 and n_2. In case B, at right, the taxonomic constitution of the original populations is different but common taxa, expressed by overlap, exist. It is worth noting that the values of C, n_1 and n_2 can be held constant while the values of N_1 and N_2 and their degree of taxonomic overlap can be varied.

Source: Henderson and Heron (1977). Reproduced with permission of the International Paleontological Association.

$$C = n_1 n_2 V / N_1 N_2 \tag{6}$$

predicts C by simple probability assuming that all samples are drawn on a random basis. The value of V, the real number of common taxa between the original populations, can be extracted from equation (6) but statistical fluctuations of C are not taken into account.

As discussed above, assessment of similarity between areas starts from binary similarity coefficients whose nature influences the final patterns obtained by multivariate techniques. In palaeobiogeographic binary data the absence of a taxon may be due to either the constitution of the original population or the differences between the latter and the sample drawn from it. The binary data specifies a sample of a population: similarity coefficients are statistical and not parametric.

The Jaccard coefficient is expressed by equation (1). Consider the case where $n_2 = N_2 = N_1$. Fig. 4.8A shows that in this case C values will equal n_1 and the value of the coefficient will vary as the ratio of n_1 to n_2; samples drawn from the same population will give widely different coefficient values depending on how n_1 and n_2 approach $N_1 = N_2$.

The Simpson coefficient is expressed by equation (2), with $n_1 < n_2$. Its value is dependent on the values of N_1, V, n_2 and N_2 as Henderson and Heron have demonstrated by substitution from equation (6):

$$S = C/n_1 = n_2V/N_1N_2 \tag{7}$$

In case of exact identity of the original population $N_1 = N_2 = V = N$, $S = n_2/N$. Thus the Simpson coefficient is controlled by the size of the larger sample (see Fig. 4.8A). The values of C are not fixed relative to the sample size (n_1 and n_2) and their probability distribution is dependent on the nature of selection of n_1n_2 from N_1 and N_2.

Henderson and Heron (1977, p. 5) concluded that:

1 similarity coefficients are inappropriate statistical measures of similarity and their application in palaeobiogeography is likely to result in misleading conclusions.
2 accordingly, multivariate analyses based on similarity coefficients are suspect as quantitative techniques.

They proposed to express similarity assessment of populations as a problem in probability, deriving a probability frequency function which relates C to n, and inferred N for each pair of OPGUs. For any pair of OPGUs, n_1 and n_2 are known; the probability distribution of C can be calculated assuming that n_1 and n_2 were drawn randomly from their original populations N_1 and N_2 whose taxonomic content is assumed to be identical. This null hypothesis is evaluated by comparing the observed value of C to the calculated probability distribution of C. The method requires random selection of taxonomic samples from larger assemblages which occur in the OPGUs. The FORTRAN program developed by Henderson and Heron takes as input data the observed n and simulated N values for each OPGU, a binary presence/absence matrix and the observed values of C for each pair of OPGUs. It calculates the probability frequency distribution for the larger N value in each pair of sites and locates the observed C value on the probability curve. The evaluation of each pair is repeated until most of the erroneous N values are adjusted, and an internally consistent suite of population diversity values is obtained. Belasky (1992) has tested the procedure by simulating two datasets and concluded that the probabilistic method was successful in substantially reducing the sampling bias in all tests on simulated data. An important result of Belasky's work is that the method provides substantially more accurate results for datasets in which taxonomic overlap exceeds 25%. This overlap can be easily obtained from data on fossil occurrence and therefore the applicability of the probabilistic method to a particular dataset can be assessed prior to analysis.

This method has not been frequently employed after the original proposal, probably because it required, at that time, excessively large computer resources (Koch, 1987). Belasky (1992, 1996) and Belasky and Runnegar (1993) have used it for biogeographic problems of the Indo-Pacific areas. An application for the estimation of palaeolongitude is presented in Chapter 8.

4.5 Are biotic provinces objective?

The most common concept of biogeographic unit is based on the definition of boundaries of areas characterized by patterns of occurrence of endemic elements. The large majority of biogeographic units are designated by 'key taxa', which are subjectively selected. Definition of boundaries is usually arbitrary and subjective because of the existence of overlapping areas or other possible solutions. Furthermore, there is a considerable diversity in concepts, definitions and denomination of the most used biogeographic units. Neobiogeographers and palaeobiogeographers use different terms and different authors belonging to the same category (neo or palaeo) may use different criteria to define the same kind of biogeographic units. For example, McKerrow and Cocks (1986) proposed to use the term 'province' for regions separated by physical barriers to migration and 'realm' for climatically controlled faunas and floras. Valentine (1973, p. 338) regarded provinces in two ways: one is as an association of taxa that must be explained in terms of the history and dispersal of their lineages and the limiting factors associated with their provincial occurrences; the other is to consider the functional aspects of the provincial system and the factors that regulate them (Valentine's provincial ecology). Palaeobiogeography is also faced with nomenclature problems caused by fluctuations of boundaries through time.

Bioprovinces are also based on assessment of similarity between OPGUs, which is mostly based on binary similarity coefficients and their treatment by multivariate techniques of analysis. But in palaeobiogeography these coefficients have been questioned and probabilistic methods are recommended.

The lack of objective criteria and the different ways of defining biogeographic units have led some authors to suggest abandoning the search for biogeographical units (Nelson, 1983; Rosen, 1984; Rosen and Smith, 1988; Smith, 1988, 1992) in favour of other methods, for example the search for 'sister areas' on the base of hierarchy of endemicities. Furthermore, nested patterns based on taxa shared between two or more areas are preferable to counts of the number of endemic taxa to one region as a measure of its degree of isolation, because endemicity may be an artefact of sampling failure or simply due to the incomplete nature of the fossil record. Even from relatively well-sampled areas a proportion of the fauna may seem endemic because it is unrecorded from elsewhere (Smith, 1988). Authors working with cladistic methods reject biotic provinces because they cannot be objectively identified on the basis of subjectively selected 'key' taxa and cannot help to develop historical hypotheses. Westermann (2000a) argued that even these more rigorous methods are based on the more or less subjective fossil taxa of the various authors.

Faunal provinces are defined by qualitative or quantitative techniques on the basis of similarities of their constituent taxa. The phyletic relations among taxa are not taken into account despite their potential biogeographical signature. Phylogenesis/area relationship introduces a historical perspective in biogeographical analysis which is discussed in the next chapter.

5 Analytical methods of historical palaeobiogeography

Several methods were devised by different authors during the 1970s and 1980s to generate historical explanations of the distributions of related taxa in different areas. These methods are mostly based on cladistic techniques of phylogenetic analysis applied here to the study of the relations between taxa and areas. The description of the various techniques is beyond the scope of this book: they are briefly summarized, focusing on the applications to palaeobiogeography of marine invertebrates. According to Wiley (1988), these techniques may be grouped into three main categories:

- area cladogram methods (Rosen, 1979; Wiley, 1980)
- consensus methods (Nelson and Platnick, 1981; Humphries and Parenti, 1986)
- parsimony methods (Brooks, 1981, 1985).

Parsimony methods have been independently developed further in two distinct types of analyses by Wiley (1988) on one hand and by Rosen and Smith (1988) on the other. These have been called respectively Brooks' parsimony analysis (BPA) and parsimony analysis of endemicity (PAE). Lieberman and Eldredge (1996) and Lieberman (1997, 2000) have proposed a different version of BPA for palaeobiogeographical applications. PAE can be applied to groups of unspecified phylogenetic relationships, whereas BPA (and the other cladistic methods) requires monophyletic groups.

5.1 The method of cladistic biogeography

Cladistic biogeography combines phylogeny and distribution of taxa by applying the techniques of cladistic taxonomic analysis to biogeographic problems (Platnick and Nelson, 1978; Nelson and Platnick,

1980, 1981; Humphries and Parenti, 1986) in an attempt to determine the history of geographic vicariance based on the history of phyletic divergence of taxa. The basic assumption is that the history of the distribution of taxa covaries with their history of speciation. The method involves the finding of **monophyletic groups** and searches for evidence of area relationships by examining the phylogenetic relationships, established by means of cladistic analyses, among taxa that inhabit the areas in question. This method produces for each clade a hierarchical nesting of sample areas through phylogenetic relationships of their endemic taxa that can be interpreted in terms of the history of the involvement of geographic areas in speciation events, i.e. the historical biogeography of the group. Although the idea of origination as a truly biogeographical process is hypothetical, as a pattern it has an empirical basis in the form of an area characterized by the most primitive taxa within a clade.

On the basis of the geographic distribution of the species in a single cladogram, the cladograms of other taxa can be examined to determine the extent to which geographically consistent patterns emerge. If the two most closely related species are invariably found in the same two areas, the geographically congruent cladograms can be used to produce **area cladograms**. These are produced for the taxa whose names at the terminal tips of the cladograms are replaced by the names of areas in which each taxon occurs (Fig. 5.1). Area cladograms are statements about relationships, not statements about centres of origin, dispersal, vicariance events, therefore a variety of interpretations of the biogeographic history of the areas considered in a cladogram is possible. It is noteworthy that dispersalist explanations are not refused by some cladistic biogeographers (Cracraft, 1980; Brundin, 1988; Wiley, 1988). **Congruence** between the area cladograms of different monophyletic groups is taken to be evidence for large-scale events and also to reconstruct palaeogeographies (Lieberman, 2000). Component analysis using consensus techniques for searching congruence has been proposed by Nelson and Platnick (1981). The summary of the main points, listed below, is taken from Patterson's (1983) review.

1 Biogeographical patterns are comparative and must include a minimum of **three** geographic components (as for statements of systematic relationships).
2 The smallest unit of biogeographical pattern is an area occupied by at least one endemic species.
3 Widely distributed species provide ambiguous information on biogeographical pattern, similarly to symplesiomorphic characters in

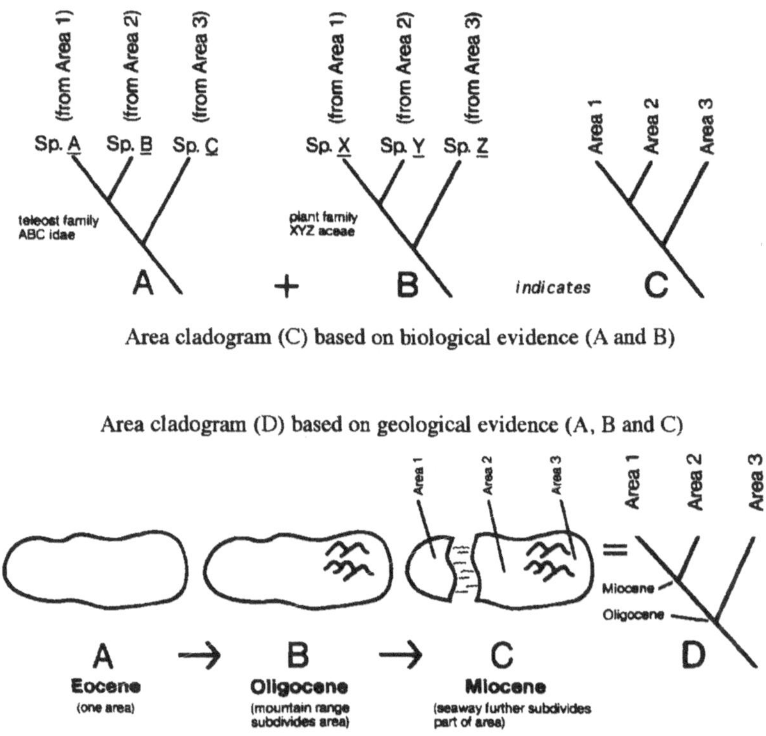

Figure 5.1 Area cladograms based on biological and geological evidence.
Source: Grande (1985). Modified with permission of the Paleontological Society.

systematics, but may contribute to the pattern of an area of endemism located within the limits of their distributional areas.

4 Species cladograms of different, partially or completely sympatric taxa may be combined into a general cladogram of areas: widespread species occurring in more than one of the areas, species cladograms which lack representatives in one or more areas ('missing taxa'), and also those with more than one species endemic to an area, may all be combined in area cladograms.

5 A species cladogram missing one or more areas, or including widespread species, or with redundant species, or with any two of these, or with all three, will specify some subset of the set of all possible cladograms for the areas concerned. Combining several such species cladograms may allow one to select the one or more members of the total set which are common to all the subsets specified.

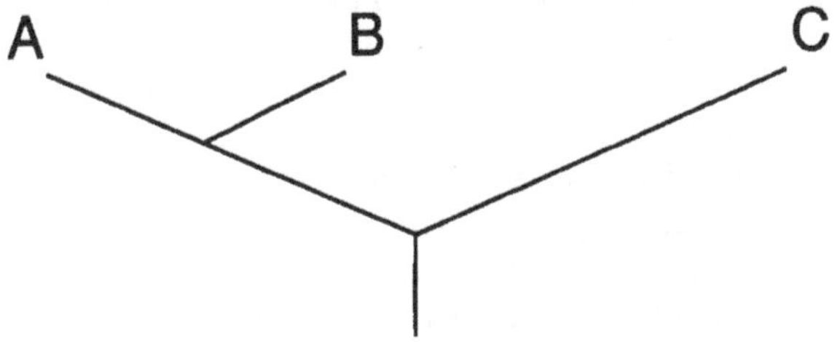

Figure 5.2 Cladogram showing relationships between three taxa. The ancestor of A + B is younger than that of A + B + C.

6 Alternative methods of dealing with widespread species, missing areas and redundant representation of areas in species cladograms are Nelson and Platnick's 'assumption 1' and 'assumption 2'. Assumption 1 assumes that both entities have the same relationships to other members of their group, whereas in assumption 2 the two entities do not have the same relationship among themselves relative to other members of the group. The latter assumption is considered as the more realistic of the two (Patterson, 1983).

As regards point (1), some further comments are needed. In phylogenetic systematics, a cladogram is informative only if a minimum of three taxa is taken into account. In fact, if only two taxa are considered in the analysis, then, though they are related, the analysis could not show whether this relationship is very distant or very close. Conversely, if we are dealing with three taxa, two of them are always more strictly related to each other than to the third one. The cladogram of Fig. 5.2 shows that the ancestor of A + B is younger than that of A + B + C. The same 'rule of the three taxa' must be applied to palaeobiogeographic analyses. The cladistic analysis allows one to state which of the three taxa are the most closely related between them and then to draw conclusions about the relations among the three areas.

Wiley (1988) pointed out some problems of the component analysis, particularly the problem of the 'missing taxa' (points 4 and 5) and the problem of the 'widespread species' (points 3 and 6). Possible explanations of the absence of a taxon in an area might be:

1 its ancestor was never in the area and no descendant of that ancestor ever dispersed into the area
2 extinction of the taxon's representatives in the area
3 sampling deficiencies.

Component analysis adds the missing area to the area cladogram of a taxon, but 'Adding an area to a cladogram of a group not known to occur in that area is analogous to claiming that a taxon has a character that has not yet been observed' (Wiley, 1988, p. 274).

The treatment of the widespread species, i.e. species distributed in more than one area, requires a number of extra operations though, according to Wiley (1988), this problem is overcome by using parsimony techniques. Widespread species provide ambiguous information because two interpretations are possible:

1 a species did not respond to a vicariance event, leaving it widespread in areas that are adjacent to each other on the area cladogram
2 a species dispersed into an area, leaving it widespread in areas either adjacent or separated by other areas on the area cladogram.

This problem is not applicable to monophyletic groups but only to single species. This must be kept in mind when we are dealing with analysis treating genera instead of species, a widespread practice in palaeobiogeography. In any case, component analysis deals with widespread species by invoking the two assumptions of point 6 above, analytically considering a widespread species as two entities. This has been strongly criticized by advocates of parsimony analyses (Wiley, 1988 and references therein).

Different equally well-supported trees are combined in a single tree by consensus techniques, a method to evaluate how the different trees agree. The use of these techniques in phylogenetic studies has been criticized because the best-supported, most parsimonious pattern is often not recovered. Furthermore, consensus techniques synthesize the results from the cladograms for each of the different groups, giving an equal weight to each one. This may cause loss of resolution in the consensus tree. For these reasons Lieberman (2000) has stressed that consensus techniques should be used with caution in biogeographical analyses.

5.1.1 Cladistic techniques and palaeobiogeography

The impact of cladistic analyses on palaeobiogeography of marine invertebrates has been minimal. This could be explained by the lack of congruence among competing cladograms or, as Jablonski *et al.* (1985) suppose, by the reluctance of palaeontologists to generate cladograms for fossil species. This method requires good cladistics leading to the recognition of monophyletic groups, and the state of taxonomy in many fossil groups precludes its use. In any case, the cladistic biogeography has certainly introduced a new scientific rigour, emphasizing the testing of models and discouraging *ad hoc* hypotheses. Its great potential is the explicitness and clarity in which biogeographic hypotheses are expressed (Jablonski *et al.*, 1985).

Hallam (1994) cautioned that area cladograms are not derived from the explicit analysis of geographic or geologic characters analogously to the way synapomorphies are analysed in systematics. Thus, whatever the superiority of cladistics for deriving phylogenies and historical hypotheses about biotas, problems remain as to how these can be converted into geological hypotheses. Furthermore, the method can only generate hypotheses of area and biotic divergence, while palaeontologists know that biogeographic convergence also occurred (Rosen, 1984, 1988c).

5.1.2 Cladistic biogeography and the use of fossil data

Because of the incompleteness of the fossil record, phylogenetic and biogeographic use of fossils has been questioned by vicariance biogeographers who use the cladistic approach (Cracraft, 1980; Patterson, 1981, 1983). Phylogenetic schemes and phylogenetically-derived biogeographical hypotheses assuming the fossil record to be complete have been considered suspect. According to the cladistic logic, fossils cannot falsify a phylogeny which is based on living taxa. This is a crucial point, because supposed fossil ancestors are often used as evidence for 'dispersal' explanations.

Despite the obvious problems of the fossil record that constrain the nature of the palaeobiogeographical data, the inclination to be dismissive of fossil information would lead to a considerable loss of information. Fossils provide information of primary importance (Grande, 1985) and can be used to derive testable biogeographical hypotheses (Rosen, 1992; Hallam, 1994). In particular, the synoptic comparison of fossil distributions over successive time intervals is unique to palaeobiogeography because it allows timing and rates for events and processes to be established. The cladistic approach of vicariance neobiogeographers applied to distributions of living organisms provides minimum ages of divergent events only if fossil data are incorporated in the analysis. Furthermore, very few extant taxa of a taxonomic level lower than class or phylum have good fossil records older than the Mesozoic, and the extinct taxa are progressively more important in older geological periods. Therefore the palaeobiogeographical methods have a value for understanding Palaeozoic (i.e. half of the duration of the Phanerozoic Eon) events, even at order and family levels.

The distributional patterns of some typical communities (e.g. reefal communities) is clearly controlled by climate and/or latitude. This provides, for the fossil record, palaeolatitudinal information for palaeogeographical reconstructions which can be tested through comparison of reconstructions derived from different methods and data sets, and also allows palaeobiogeographical hypotheses to be generated.

The fossil record of marine invertebrates is incomplete but certainly less lacunous, by far, than that of terrestrial organisms and in some cases it allows detailed phylogenetic analyses (ammonites, trilobites, planktonic foraminifers, rudists, etc.). These can be used, with a certain degree of confidence, for palaeobiogeographical purposes.

Finally, Benton *et al.* (2000) have recently demonstrated the congruence between the order inherent in the evolutionary trees (phylogeny) and the order of fossils in the rocks (stratigraphy). This shows no evidence of diminution of the quality of the fossil record backwards in geological time, even though ancient rocks clearly preserve less information, on average, than more recent ones.

5.2 Brooks' parsimony analysis (BPA)

Brooks (1981, 1985, 1990) proposed a parsimony technique (originally devised for analysing host phylogenies by using the phylogenies of their parasites) for examining the degree to which, and the specific points in which, the phylogenetic diversification of geographically and ecologically associated clades coincided (co-speciation). Brooks' technique treats the problem as an area/taxon matrix problem in which areas and taxa are respectively equated with taxa and characters of the cladistic phylogenetic analysis. This parsimony technique was examined in detail by different authors, particularly by Wiley (1988 and references therein), who termed it 'Brooks parsimony analysis' (BPA). Emphasis was given to this method because it does not simply address the problem of area relationship but focuses on coevolutionary patterns among clades or parts of clades that inhabit the same areas. It does not necessarily depend on the *a priori* identification of areas of endemism or biotas. On the other hand, BPA offers solutions to problems encountered by Nelson and Platnick's (1981) component analysis using consensus techniques, namely the treatment of widespread species, missing taxa and incongruent area cladograms.

In practice, the first step is the reconstruction of the phylogenetic relationships of the members of the clade through cladistic analysis. The second step is to designate the areas of occurrence of the species

Table 5.1 Area/taxon matrix listing binary codes that represent the geographical distribution of hypothetical species. These represent biogeographic data (Brooks, 1990).

Area	*Sp. 1*	*Sp. 2*	*Sp. 3*	*Sp. 4*	*Sp. 5*	*Sp. 6*	*Sp. 7*	*Sp. 8*	*Sp. 9*
North America	1	0	0	0	0	1	0	0	1
Europe	0	1	0	0	0	1	0	0	1
Australia	0	0	1	0	0	0	0	1	1
South America	0	0	0	1	0	0	1	1	1
Africa	0	0	0	0	1	0	1	1	1

(as if they were taxa) and to list the geographic distributions of the different species involved in the analysis: presence is listed as '1' and absence as '0' (Table 5.1). Terminal species are coded as 'apomorphies' for the areas in which they occur, while hypothetical ancestral species are coded as 'synapomorphies' uniting areas. The matrix is then processed with appropriate computer programs which generate trees.

Parsimony analysis yields a consistency index (CI) for each tree, which gives a measure of how often each variant (taxon) appears in the calculation. In other words, it provides a measure of confidence in the possibility of other, equally parsimonious trees, i.e. cladograms, existing besides the one choosen. A CI of 1.0 corresponds to a whole congruence of the data with the result. A low CI (less than 0.3, for example) indicates that little confidence can be placed on a given tree topology. Conversely, a higher CI indicates greater confidence.

Finally, an area cladogram based on the phylogenetic relationships of the species is realized to represent the historical involvement of areas in species evolution. This can be compared with geological data to assess congruence (Fig. 5.3).

Note the numbers accompanying slash marks in the cladogram. They indicate phylogenetic relationships of the five hypothetical terminal species 1–5: each taxon and each internal branch of the tree is numbered: each species has a 'code' that indicates both its identity and its common ancestry. For example, the code for species 1 (1,6,9) indicates that 9 is the common ancestor of all species (it occurs in all areas) and 6 is the ancestor of both species 1 and 2.

Problems might result from parsimony analysis, for example ancestors appearing at two different internodes or an ancestor appearing with its own descendant within the same area, implying that the initial coding was 'wrong', i.e. *a posteriori* analysis leads one to such a conclusion. Other problems can be due to the misplacement of ancestors

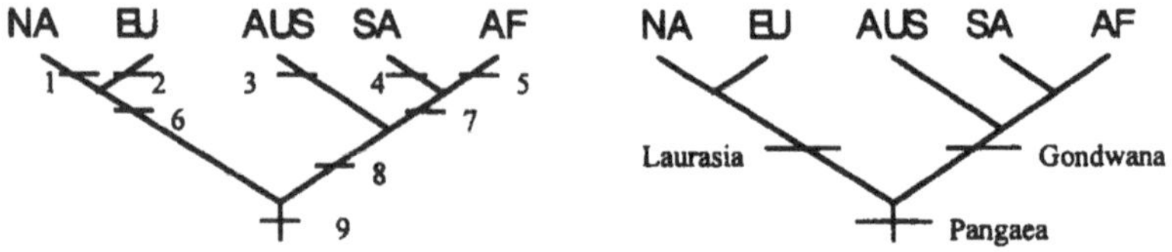

NA=North America; EU=Europe; AUS=Australia; SA=South America; AF=Africa

Figure 5.3 Left: area cladogram for five areas based on the phylogenetic relationships of five hypothetical terminal species that inhabit those areas; numbers accompanying slash marks indicate codes for species as listed in Table 5.1. Characters (species) 1 to 5 are terminal taxa whereas 9, 8, 7 and 6 correspond to nodes. Right: area cladogram based on historical geological data.

Source: Brooks (1990). Modified with permission of the editor.

on the internodes of the area cladogram. In all these cases the area cladograms may be controlled by means of the phylogenetic relationships established with cladistic analysis.

It is impossible here to discuss at length all the analytical and interpretative solutions to problems of missing taxa, widespread taxa or more than one member of the clade in the same area discussed by Brooks (1990). However, the problem of missing taxa deserves a comment because of its relevance to biogeography. In many cases not all representatives of a group are represented in all areas under study. Neobiogeographers are unable to state *a priori* whether extinction explains the absence of a taxon from an area or, conversely, whether dispersal (with or without speciation) is one explanation for representatives of certain clades having been added to an area. In these cases missing taxa can be coded with a question mark, '?', in the data matrix. On the other hand, palaeontologists may have arguments for considering a taxon really absent because of extinction and therefore for coding as really absent with '0'.

Wiley (1988) proposed a version of BPA that uses rooting by introduction of an ancestral vector of zero values. This represents biogeographic characters analogous to characters assigned to the outgroup node in a phylogenetic analysis, i.e. that any distributional data prior to the origin of each group are considered plesiomorphic or irrelevant.

For a recent, and detailed, review of BPA I refer the reader to Lieberman (2000). Box 5.1 summarizes the palaeontological use of BPA.

Box 5.1 Palaeontological use of BPA

Although BPA was introduced by neobiogeographers, it has recently been used to recognize palaeobiogeographical patterns inherent in previously established phylogenies of Middle Devonian trilobites by Lieberman and Eldredge (1996). The method was used to determine the relative prevalence of vicariant patterns versus repeated patterns of dispersal in several trilobite clades. Lieberman (1997) also performed BPA for Cambrian trilobite palaeobiogeography. Wiley's (1988) 'outgroup version' of BPA has been used in three separate analyses with the biogeographic states from trilobite phylogenies coded into three distinct matrices in order to distinguish between episodes of vicariance and episodes of 'geo-dispersal' (Lieberman, 2000).

The first analysis treated a data matrix where only the biogeographic states of nodes and terminal taxa that fit with the vicariant pattern of progressive range contraction and diversification have been coded. In the second analysis only the biogeographic states of nodes and terminals that fit a pattern of dispersal (omitted from the first one) were coded in the matrix. These data matrices were finally combined to obtain a complete biogeographic picture using both vicariance and dispersal data. For each analysis the data matrices were subjected to a parsimony analysis using the software PAUP (Swofford, 1990) and all nodes have been optimized using Fitch's (1971) optimization which allows transitions to occur between all biogeographic states in all possible directions so that all transformations are unordered. This optimization is a good technique when patterns of area relationship are unknown because it does not assume *a priori* the direction of biogeographic change.

In order to show how the analysis has been carried out, the manner in which data have been coded is summarized here taking the example of one of the five clades used by Lieberman and Eldredge (1996).

Figure 5.4 presents the phylogeny of the trilobite genus *Basidechenella* with biogeographic states substitued for terminal taxa and mapped onto the ancestral nodes. Tables 5.2 and 5.3 show the biogeographic character states coded in two distinct matrices for vicariance and for dispersal analyses respectively. Seven areas plus the 'ancestor' or 'outgroup area' have been considered. The data matrices show that in its ancestral palaeo-biogeographic state

Box 5.1 continued

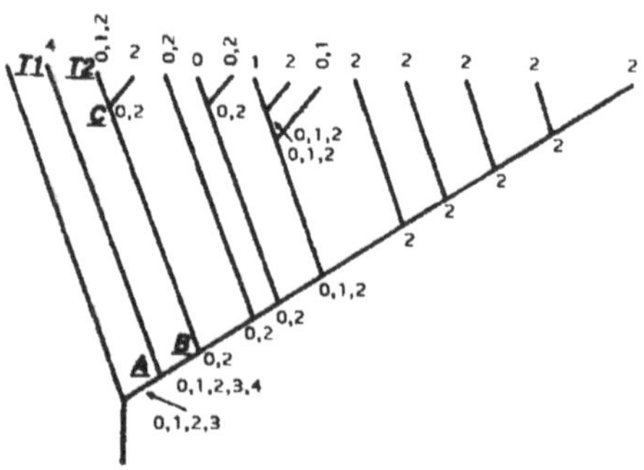

Figure 5.4 Phylogeny of the trilobite genus *Basidechenella* (from Lieberman and Eldredge, 1996). Numbers correspond to areas: 0, Appalachian Basin; 1, Illinois Basin; 2, Michigan Basin; 3, Arctic; 4, Armorica; 5, Kazakhstan; 6, N. South America. Nodes and terminal taxa correspond to distinct trilobite species.

Source: Lieberman and Eldredge (1996). Reproduced with permission of the Paleontological Society.

Basidechenella is present only in areas 0, 1, 2 and 3, respectively Appalachian Basin, Illinois Basin, Michigan Basin and Arctic. The absence of a taxon is treated as primitive and considered to represent true absence, and listed as '0' in the data matrix, while the presence is treated as derived and listed as '1'.

In the **vicariance analysis** the transition from the ancestral biogeographic state to Node A (= 2 in Table 5.2), which is a taxon also present in area 4 (Armorica), represents a range expansion (treated by Lieberman and Eldredge as 'dispersal') and it is **not** coded as an ordered multi-state character in the vicariance analysis. Conversely, range contraction is represented by the transition from node A to the terminal taxon T1 (= 3 in Table 5.2) because the latter is present only in area 4. This transition, interpreted as vicariance, is represented using taxon T1 as an ordered multi-state character which is coded as '1' (i.e. present even if it is actually absent) in areas 0, 1, 2 and 3, with a special derived presence in area 4 where it is coded as '2'. Another episode of range contraction and vicariance is also marked by the transition between node B (= 4 in Table 5.2) and its ancestral node (node A) because its biogeographic state signifies

Box 5.1 continued

Table 5.2 Vicariance analysis, modified from Lieberman and Eldredge (1996).

Area	*1*	*2*	*3*	*4*	*5*	*6*	*7*	*8*	*9*	*10*	*11*	*12*	*13*	*14*
'Ancestor'	0	0	0	0	0	0	0	0	0	0	0	0	0	0
0 Appalachian Basin	1	1	1	2	1	1	1	1	1	1	1	1	2	1
1 Illinois Basin	1	1	1	1	0	1	0	0	0	0	0	0	0	1
2 Michigan Basin	1	1	1	2	1	1	2	1	1	1	1	1	1	1
3 Arctic	1	1	1	1	0	0	0	0	0	0	0	0	0	0
4 Armorica	0	1	2	1	0	0	0	0	0	0	0	0	0	0
5 Kazakhstan	0	0	0	0	0	0	0	0	0	0	0	0	0	0
6 N. South America	0	0	0	0	0	0	0	0	0	0	0	0	0	0

Area	*15*	*16*	*17*	*18*	*19*	*20*	*21*	*22*	*23*	*24*	*25*	*26*	*27*	*28*
'Ancestor'	0	0	0	0	0	0	0	0	0	0	0	0	0	0
0 Appalachian Basin	1	1	2	1	1	1	0	0	0	0	0	0	0	0
1 Illinois Basin	1	1	2	2	1	1	0	0	0	0	0	0	0	0
2 Michigan Basin	1	1	1	1	2	2	1	1	1	1	1	1	1	1
3 Arctic	0	0	0	0	0	0	0	0	0	0	0	0	0	0
4 Armorica	0	0	0	0	0	0	0	0	0	0	0	0	0	0
5 Kazakhstan	0	0	0	0	0	0	0	0	0	0	0	0	0	0
6 N. South America	0	0	0	0	0	0	0	0	0	0	0	0	0	0

Table 5.3 Dispersal analysis, modified from Lieberman and Eldredge (1996).

Area	*1*	*2*	*3*	*4*	*5*	*6*	*7*	*8*	*9*	*10*	*11*	*12*	*13*	*14*
'Ancestor'	0	0	0	0	0	0	0	0	0	0	0	0	0	0
0 Appalachian Basin	1	1	0	1	1	1	0	1	1	1	1	1	1	1
1 Illinois Basin	1	1	0	0	0	2	0	0	0	0	0	0	0	2
2 Michigan Basin	1	1	0	1	1	1	1	1	1	1	1	0	1	1
3 Arctic	1	1	0	0	0	0	0	0	0	0	0	0	0	0
4 Armorica	0	2	1	0	0	0	0	0	0	0	0	0	0	0
5 Kazakhstan	0	0	0	0	0	0	0	0	0	0	0	0	0	0
6 N. South America	0	0	0	0	0	0	0	0	0	0	0	0	0	0

Area	*15*	*16*	*17*	*18*	*19*	*20*	*21*	*22*	*23*	*24*	*25*	*26*	*27*	*28*
'Ancestor'	0	0	0	0	0	0	0	0	0	0	0	0	0	0
0 Appalachian Basin	1	1	1	0	0	0	0	0	0	0	0	0	0	0
1 Illinois Basin	1	1	1	0	1	0	0	0	0	0	0	0	0	0
2 Michigan Basin	1	0	1	1	0	1	1	1	1	1	1	1	1	1
3 Arctic	0	0	0	0	0	0	0	0	0	0	0	0	0	0
4 Armorica	0	0	0	0	0	0	0	0	0	0	0	0	0	0
5 Kazakhstan	0	0	0	0	0	0	0	0	0	0	0	0	0	0
6 N. South America	0	0	0	0	0	0	0	0	0	0	0	0	0	0

Box 5.1 continued

presence in areas 0 and 2 only. Thus, character 4 is coded as an ordered multi-state character: its real presence in areas 0 and 2 is coded as '2' and its absence in areas 1, 3 and 4 as '1', the absence in areas 5 and 6 being coded as '0'. No change in geographic range is involved in the transition from node B to node C (= species 5) whose biogeographic state is represented by coding absence as '0' and presence as '1'. Range expansion is involved in the transition from node C to the terminal taxon T2, which is present in areas 0, 1 and 2. This does not conform to a pattern of vicariant splitting and is treated as an unordered multi-state character: presence is coded as '1' and absence as '0'.

In the **dispersal analysis** range expansions have been treated as ordered multi-state characters. Repeated patterns of range expansion into the same area were interpreted as evidence for 'geo-dispersal'. The trans-ition from the ancestral biogeographic state to node A (Fig. 5.4) marks an episode of dispersal; therefore, contrary to vicariance analysis, character 2 (= node A) is treated as an ordered multi-state character and its presence in area 4 is coded as '2' in the matrix of Table 5.3, while in the vicariance analysis it was coded as '1'. Conversely, the transition from node A to terminal taxon T1 marks a range contraction interpreted as an episode of vicariance. Accordingly, in the dispersal analysis this transition is **not** coded as an ordered multi-state character and its presence in area 4 is coded as '1' while its absence in the other areas is coded as '0'.

The results of both types of analysis, which have been performed on large data matrices (168 codings from the phylogenies of five clades of trilobites), are shown by the cladograms of Fig. 5.5. The most parsimonious tree produced by parsimony analysis of the matrix created for vicaricance analysis shows (Fig. 5.5A) two 'monophyletic' groups of areas: Armorica and Arctic (respectively areas 3 and 4 in the matrix) are sister areas as well as Michigan and Appalachian basins (respectively areas 0 and 2), which form a monophyletic group with Illinois basin (area 1). Due to high values of both CI and r.i. (retention index), Lieberman and Eldredge conclude that the vicariance data have a strong biogeographical signal. Bootstrap analysis, conducted to assess the relative support for each of the nodes of the tree, produced high confidence values for the same areas. According to Lieberman and Eldredge, a strong

Box 5.1 continued

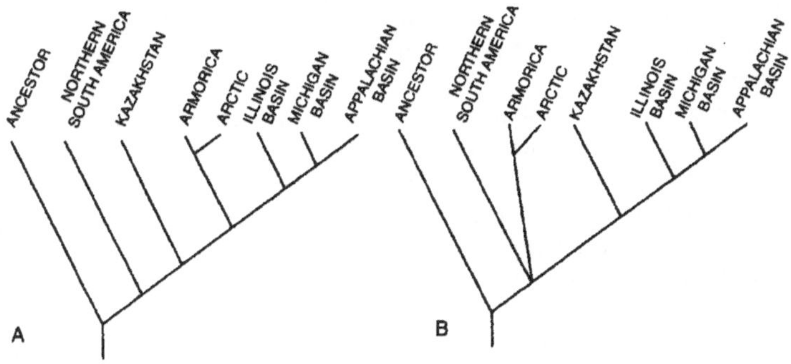

Figure 5.5 A: the most parsimonious cladogram produced from the matrix for vicariance analysis showing the pattern of area relationship. Length is 207 steps, r.i. is 0.74 and CI 0.85 when uninformative characters are excluded. B: strict consensus cladogram of the three most parsimonious trees produced by the matrix for dispersal analysis showing the pattern of area relationship. Length is 207 steps, r.i. is 0.66 and CI 0.78 when uninformative characters are excluded.

Source: Lieberman and Eldredge (1996). Reproduced by permission of the Paleontological Society.

biogeographic signal is also contained in the dispersal data because the three most parsimonious trees produced after parsimony analysis are characterized by high values of CI and r.i. The strict consensus of the three most parsimonious trees is shown in Fig. 5.5B. The same relations exist between Illinois, Michigan and Appalachian basins, although here Kazakhstan (area 5 in the matrix) is their sister area, and Arctic and Armorican sister areas form part of a basal polytomy along with northern South America (area 6 in the matrix). Results of the bootstrap analysis also give a strong support for close area relationships between the three eastern North American basins but moderately strong support for area relationships between Arctic and Armorica. Dispersal and vicariance data matrices were finally combined and subjected to parsimony analysis producing three most parsimonious trees of 510 steps, with CI 0.78 and r.i. 0.71. A strict consensus of these trees has the same topology as the tree produced by the dispersal analysis

Box 5.1 continued

(Fig. 5.5B) and bootstrap strongly supported area relationship between the eastern North American basins and between Arctica and Armorica.

The biogeographical interpretation of the results of these analyses takes into account the identical patterns produced by dispersal and combined analyses (Fig. 5.5B), which indicate that the distributional data governed by dispersal predominate over those governed by vicariance. This implies the presence of a signal of 'geo-dispersal' in the phylogenetic and distributional data from the five clades considered. Congruence in several components of the vicariance cladogram (Fig. 5.5A) with the consensus cladograms produced in the dispersal and combined analyses suggests that the most recent dispersal history corresponds to the most recent vicariant history. For example, the final separation of the Michigan basin from the Appalachian basin is a more recent event than the progressive isolation of Michigan and Appalachian basins from Illinois basin, but cladograms from the analysis of dispersal also indicate that the closest dispersal relationship and the most recent shared dispersal history are between the Appalachian and Michigan basins. The role of sea-level changes has been invoked for dispersal pattern. However, the statements 'the last barriers to come up were the first to fall, and under changing geological conditions the last areas to differentiate would be the first to merge' (Lieberman and Eldredge, 1996, p. 74) can be criticized because filter effects could control some of the vicariant and/or dispersal patterns within the studied clades. The differences between the different cladograms (Fig. 5.5A, B) indicate complex area relationships between Kazakhstan and the other regions which are interpreted as a possible artefact produced by the paucity of trilobite lineages from that region.

5.3 Parsimony analysis of endemicity (PAE)

The method of 'parsimony analysis of endemicity' (PAE) was first attempted by Rosen (1985) and realized more fully by Rosen and Smith (1988). Similarly to BPA, PAE also treats each taxon present at a sample locality analogously to a character in cladistic phylogenetic analysis. Despite the coincident structure of the data matrix, PAE differs from both cladistic biogeography and BPA in being applied to groups whose phylogenetic relationships are unspecified and which do not need to be clades of proven monophyly, though the individual taxa used should be as far as possible non-polyphyletic. This is of particular relevance for palaeobiogeographic analyses as there are few well supported phylogenies of fossil groups. PAE also differs from cladistic biogeography in using sets of contemporaneous taxa and not related taxa from any geological horizon (Rosen, 1988b).

PAE was devised in order to study patterns of endemicity through parsimony analysis of taxa shared between sample localities without reference to any geographical prejudgement, e.g. arbitrary arrangement in subjective biogeographic units like provinces. In fact, an important result of PAE is the abandonment of the search for faunal provinces, which are considered arbitrary entities based on subjectively designated endemic 'key taxa', in favour of the search for sister areas. This technique has also been used in neobiogeographical studies (see Ron, 2000, and references therein).

Like multivariate analysis, presence/absence data for a set of sample localities and a particular group of organisms are taken as starting point in PAE. The same kind of area/taxon matrix used in BPA and also in cluster analysis is used in PAE, but instead of treating presence/absence data in terms of overall similarity, it uses parsimony analysis of shared presence of taxa to identify a nested hierarchical pattern, as in cladistics. Therefore shared presences are treated analogously to synapomorphies in systematics.

Rosen (1988b) listed some guidelines to modify the raw data matrix (taxonomic lists at sample localities) prior to analysis in order to maximize the biogeographical resolution:

1 cosmopolitan taxa (actually taxa occurring in all sample localities) as well as those occurring in a single locality ('autapomorphs') are omitted as uninformative for finding relationships between the localities
2 localities with very few taxa in relation to most of the other localities should be omitted or merged because they may give equally

parsimonious results and, using the 'Lundberg rooting', their low diversity is interpreted as 'primitive'

3 localities with an identical set of taxa should be merged because they analytically represent a single region.

The most parsimonious hierarchical arrangement of sample localities is obtained through processing the data matrix with a computer program. The matrix represents the occurrence (presence/absence) of taxa with respect to a set of sample localities of the same age. The different successive versions of Swofford's PAUP have been used by the authors who dealt with PAE (Rosen and Smith, 1988; Smith, 1988, 1992; Fortey and Mellish, 1992; Waggoner, 1999); I refer here to version 3.0 (Swofford, 1990). Once the data block has been established, the assumption 'ancstates allzero' must be specified at the end of the input file and then selected in the 'Data' menu. The Lundberg rooting is used excluding uninformative characters by selecting 'Ignore uninformative' in the 'Options' menu. In analogy to phylogenetic analysis (and similarly to Wiley's version of BPA discussed above), an 'outgroup' is used to root the cladogram and polarize the character states. Such an 'outgroup' is a sample locality without any of the taxa which are present in the rest of the sample locality set, i.e. an area occurring so far back in geological time that none of the taxa in the sample set of localities had yet evolved. The alternative of using an outgroup area containing all taxa would bring to group areas according to shared absences; this implies depletion through time starting from a cosmopolitan biota. Rosen and Smith (1988) and Smith (1988) originally rooted the trees using the 'Lundberg' option, where the total absence of taxa was taken as primitive and the distribution of all shared taxa as derived.

Dichotomies in PAE cladograms represent biogeographical divergences of at least one pair of taxa between the sample localities under analysis. Dichotomies nearer the origin of the cladogram represent older divergences. Dichotomous patterns are indicative of biotic divergences but **convergence** is also a possible historical event. Because PAE nests localities according to how recently they have shared a common biota, irrespectively of possible previous successions of convergent and divergent events, analyses on different time-planes are performed. The minimum time interval between time-planes should be appropriate to the taxonomic level of the data being analysed.

It is worth noting that opposite processes such as vicariance and 'geo-dispersal', the latter being analogous to convergence, were investigated with BPA by Lieberman and Eldredge (1996) and Lieberman

(1997) by coding in two different ways a single data set corresponding to the same time-plane. In contrast, PAE can demonstrate these conceptually opposite processes through interpretation of analyses of a succession of chronologically different data sets.

The model presented by Rosen and Smith (1988) provides a hypothetical biogeographical history involving alternately convergent (i.e. barrier disruption) and divergent (i.e. vicariance) events between each time-plane. Divergent events result in two daughter biotas with a 'biotic fingerprint', i.e. a proportion of taxa in common inherited from their ancestral biota. This fingerprint increasingly weakens through time (see in Fig. 5.6 the progressive disappearance of biota A from t_1 to t_6). Convergences are not shown graphically by the cladograms but they can be inferred by comparing cladograms over a time sequence if between any pair of successive time intervals there is a transposition of sister relationships between three biotas. In Fig. 5.6 the biota D is sister to biota A_1'' at time t_4 but is sister of the C-lineage at time t_5 because of convergence at t_4/t_5. The two biotas which have intermixed are the transposed biota D (in DC_1) and its sister biota C (in C_2'), though the cladogram only shows a divergence between them. PAE provides other important elements for reconstruction of biogeographical history of biotas. For example, the cladogram at t_6 would suggest that all biotas were united at some time in the past. But the analysis at different time-planes indicates a more complex historical succession of vicariant and convergent events whose final result is the biogeographic relationship observed at t_6. The inferred history of the biotas can be used for palaeogeographical reconstructions. Cladograms only provide historical hypotheses about biotas, as represented at sample localities, in terms of recency of intermixing between these localities. Thus, area cladograms represent hypotheses of area histories. Divergences can be interpreted as hypothetical vicariance events, implying that biotic exchange has been stopped by barriers that might have appeared in regions located between pairs of sister-sample localities on each side of a divergence. However, the nature of both barriers and events concerned cannot be specified from any analysis, neither the palaeogeographical proximity of sample localities nor the effectiveness of the barriers. Biogeographical analysis should provide in fact test-like comparisons with independently derived histories. Of course, if no independently derived palaeogeographical models based on geological data are available, a biotic event inferred from biogeographical analyses can be taken at least as a starting point for further research.

PAE suffers from some problems, for example sampling bias and homoplasies. Concerning sampling bias, all absences of taxa from a

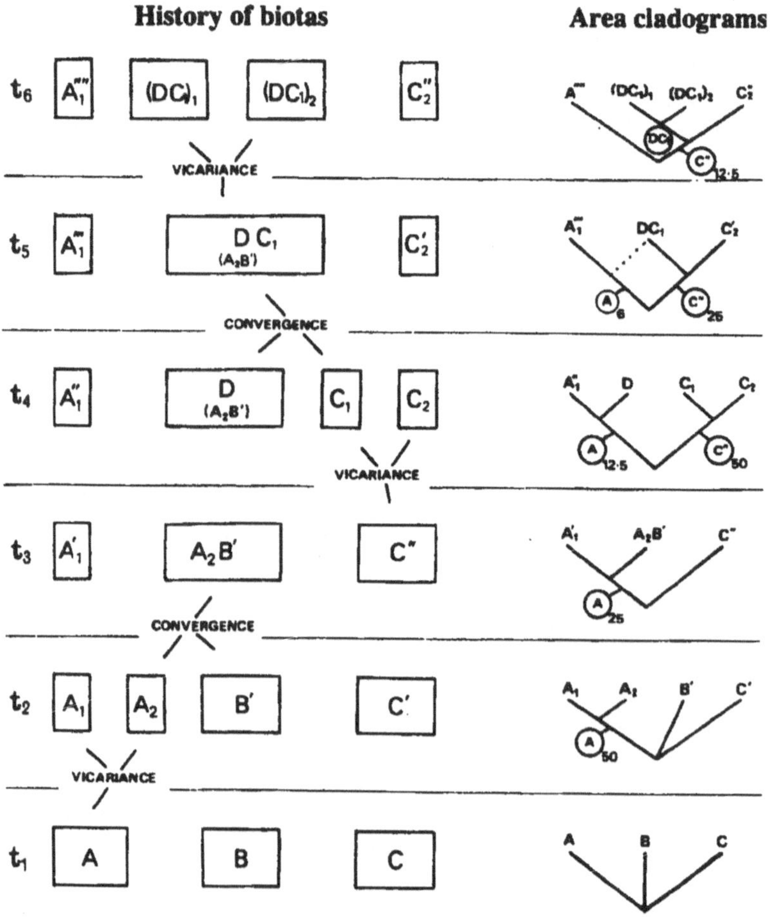

Figure 5.6 A hypothetical history of biotas and the corresponding sample locality cladograms. Rectangles correspond to geographical sample localities and letters to biotas (assumed to be fully sampled). t_1–t_6 are successive time-planes. The initial faunas A, B, C have no taxa in common. C′, C″, C‴, etc. represent turnover of biota C through time. D is an endemic biota originated by convergence. Common elements of biota shared between sister localities are circled, whereas numbers indicate for each time-plane the percentage of the total biota shared at the vicariance event that has survived to this time.

Source: Rosen and Smith (1988). Reproduced by permission of the Geological Society.

biogeographical data set should be treated as unknown character states (the taxon may be undiscovered yet) but Rosen and Smith (1988) assumed that all absences are genuine. Homoplasy in a phylogenetic (taxonomic) cladogram can be explained with parallelism or convergence and non-homology by independent evolution of the same character in different lineages or even through evolutionary loss ('reversal') and subsequent re-emergence of the character. If in PAE the 'loss' is not a conceptual problem because it corresponds to a local extinction, on the contrary the reappearance or multiple independent evolution of the same 'character' (i.e. taxon) at disjunct places and/or at different times is conceptually unacceptable.

Therefore, Dollo's Law has been taken into account to avoid the possibility of 're-evolving taxa' and calculate the most parsimonious cladogram by employing the DOLLO option (maximum parsimony without reversals) available from version 3.0 of the software PAUP (Swofford, 1990). The use of this algorithm has been criticized by Lieberman (2000) because 'it assumes that the same biogeographic state (geographic distribution) is never reoccupied by another species' and 'it seems unwise to invoke *a priori* specific assumptions about how geographic differentiation occurred' (p. 120). The alternative to Dollo parsimony, other than for truly global data sets, was proposed by Rosen (1992); it consists of interpreting homoplasies as local extinctions with subsequent re-stocking of the sample areas concerned by the same taxon or taxa at one or more subsequent times from areas outside those in the data set. This may be explained by invoking the different processes included in the category of the distributional change, but additional information about the age range and distributions of the taxa in question is needed to constrain the hypotheses.

Rosen and Smith (1988) cautioned that the use of the rooting does not eliminate the possibility that PAE cladograms may reflect either wholly or in part latitudinal or other major ecological gradients. A nested hierarchy of sample localities showing a pectinated topology can be the result of either an ecological diversity gradient or a historical sequence of vicariance events. When PAE generates pectinated patterns, ecological effects and historical factors cannot be separated. PAE cladograms can reflect in part factors such as barriers or ecological conditions and may be used as an alternative to classify biotas, although cosmopolitan and 'autoendemic' taxa are omitted prior to analysis and some taxa can occur at other localities not considered for the analysis. The historical value of PAE cladograms depends on the interpretation of the 'outgroup area'. In fact a 'Lunberg root locality' can correspond either to an area where ecological conditions were not favourable to all taxa preventing their survival there, or to one that occurs so far back in geological time relative to the sample horizon that none of the taxa considered in the analysis had yet evolved. Thus, a PAE cladogram could represent a hypothesis of contemporaneous ecological conditions implying ecological relations between sister localities or represent alternatively a historical sequence of biotic interchange or divergence. It is therefore difficult to assess the ecological or the historical value of a single PAE pattern. Ron (2000) stresses that ideal organisms for the application of PAE (and also cladistic biogeographic techniques) are those that have limited dispersal abilities and speciate in a vicariant manner.

According to Lieberman (2000), results of PAE should be treated as good first approximations because this method is based on distribution of taxa whose monophyly does not need to be proven. This author stresses that, like phenetic approaches, phylogenetic relationships provide additional data relevant to historical biogeography that PAE cannot capture (Lieberman, 2000, p. 159). This criticism is based on the assumption that only phylogenesis bears a historical signature which can be captured and interpreted through methods of cladistic biogeography. But these methods are strongly affected by the problem of the extinct taxa; furthermore, they treat related taxa which actually lived in different geological horizons. In Lieberman's applications of BPA, the likelihood of the causal relevance of different biogeographical processes such as biogeographical convergence and vicariance (both embodied in Lieberman's 'geo-dispersal') is tested by coding *ad hoc* a single data set of a time interval (e.g. Middle Devonian) which lasted some millions of years. However, we need to unfold (through the analysis of distributional patterns) the events 'compressed' in a single time-plane because what we try to investigate is a historical succession of biogeographical processes. By using sets of contemporaneous taxa, PAE describes changing relations between areas through different time horizons; the historical succession of biogeographical events may then be tested against independent lines of evidence.

The possibility offered by PAE to analyse a chronological succession of different data sets, as shown in Rosen and Smith (1988) and Smith (1988), has been poorly exploited. This may be a possible reason for the criticism expressed of this method. In other words, the number of palaeobiogeographical studies is not yet sufficient to prove that PAE fulfils the expectations (but note that this remark is also valid for other analytical methods).

Two examples taken from the literature are presented in Box 5.2 to show applications of PAE to palaeogeographic problems.

Box 5.2 Examples of applications of PAE

PAE provides palaeobiogeographic constraints or independent tests to plate tectonic reconstructions as shown in the following examples.

Tectonic history of East Asia

Smith (1988) analysed Late Carboniferous–Late Permian palaeobiogeographical patterns of rugose coral genera to provide palaeontological tests for different poorly constrained rival hypotheses about the plate tectonic reconstruction of East Asia and the history of opening and closure of the different tectonic sutures. Asia formed in fact a complex of small plates accreted together, whose boundaries correspond to ophiolitic suites representing remnants of ancient oceanic tracts (Fig. 5.7). Although the existence of a Tethyan ocean

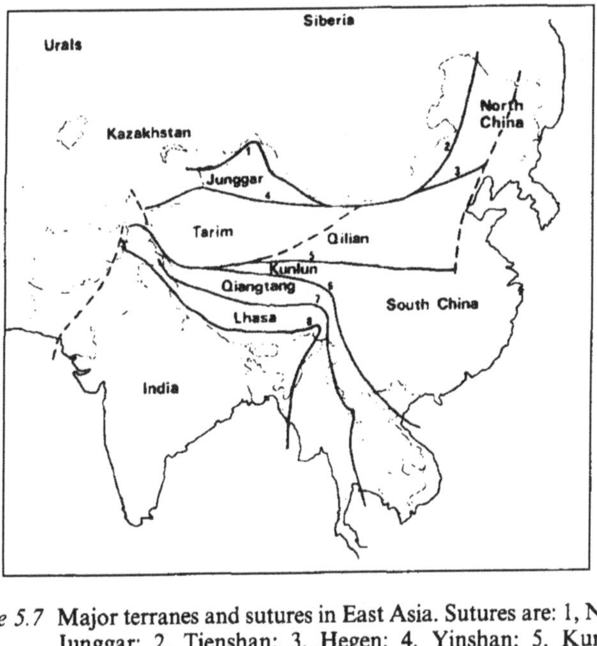

Figure 5.7 Major terranes and sutures in East Asia. Sutures are: 1, North Junggar; 2, Tienshan; 3, Hegen; 4, Yinshan; 5, Kunlun-Qinling; 6, Jinsha; 7, Banggong; 8, Zangbo.

Source: Smith (1988). Reproduced with permission of the Royal Society.

Box 5.2 continued

between Eurasia and Gondwana is generally accepted, a number of alternative hypotheses of palaeogeographical reconstruction exist. It has been supposed that a 'Palaeotethys' was separated from the 'Neotethys' by a relatively narrow continental strip, the 'Cimmerian continent', which was part of the northern margin of the Gondwana until the Permian, when it was rifted and then swept across the Palaeotethys. On the other hand, a scattering of small terranes, rather than a single Cimmerian continent, distributed across Tethys before their diachronous accretion to form present-day China and South-East Asia. Smith's study was realized to establish where and when biotic discontinuities are to be found in East Asia and to place constraints on possible plate reconstructions.

The boundaries to biotic regions were chosen to coincide with well-established tectonic lineaments (mostly ophiolitic suture zones). Principal tectonic regions were taken for this study and were subdivided in the case of significant faunal changes across a single tectonic terrane. In order to keep palaeobiogeographical history separated from any tectonic reconstruction, and therefore to use it as an independent test, Smith performed PAE for individual data sets (Fig. 5.8), comparing the results with those obtained from other numerical techniques of ordination and classification such a principal coordinates analysis (PCORDS) and single linkage cluster analysis (SLC) respectively. Empirically derived patterns based on other fossil groups were also compared to these results. Smith cautioned that the end Carboniferous glaciation must presumably have had a marked effect on patterns of distribution. Therefore the distinction between distributional changes due to shifting of climatic belts and those produced through oceanic floor spreading and plate motion is important, though this relies on narrative interpretation, incorporating other data (as facies characteristics).

The PAE cladogram for Visean shows two groupings, one uniting Kazakhstan and South Urals, the other uniting SE China, Qilian and Kunlun terranes. The remaining sample areas are arranged in a pectinate pattern of decreasing diversity. However, this pectinate pattern was subdivided by means of both PCORDS and SLC analyses showing Lhasa–Himalaya and NE China–Qiangtang groupings (Fig. 5.9).

In the Visean, the diversity of rugose coral genera is high in the SE China–Qilian–Kunlun region on one hand and in Kazakhstan

Box 5.2 continued

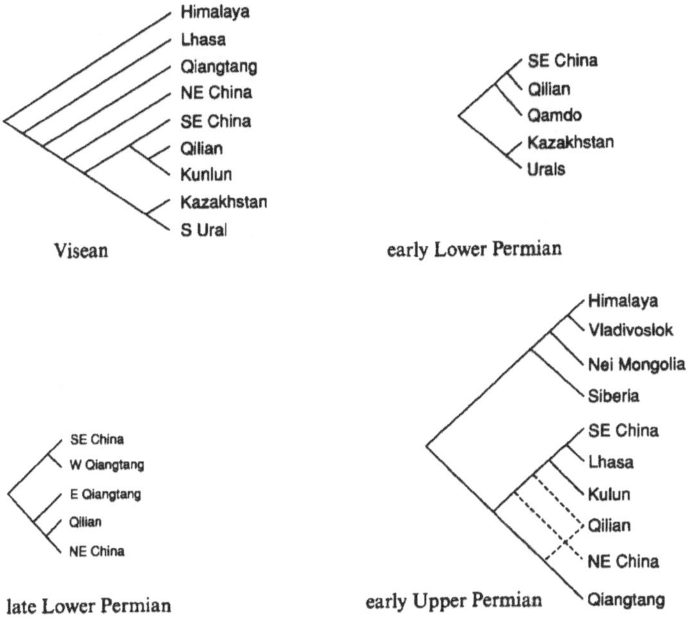

Figure 5.8 PAE area cladograms corresponding to the selected time-intervals.

Source: Smith (1988). Reproduced by permission of the Royal Society.

and the South Ural Mountains on the other. Both areas were presumably at comparable low latitudes, yet there was clearly some barrier to faunal exchange. Diversity is lower to the north in NE China and to the south in Qiangtang, Himalaya and Lhasa terranes. Faunal exchange was relatively unhinibited in the Cathaysian area (North and South China blocks, Tarim, Kunlun, Qiangtang terranes). Faunal diversity decreased southwards, towards the Himalaya, though no endemic faunas existed there. During the Early Permian, the period of extensive glaciation in southern hemisphere, coral faunas have a low diversity. The isolation of Cathaysian equatorial faunas from Central Asia is still apparent. Lhasa, Himalaya and the western part of the Qiangtang terranes presumably lay in higher latitudes than the Cathaysian region, as suggested by the

Box 5.2 continued

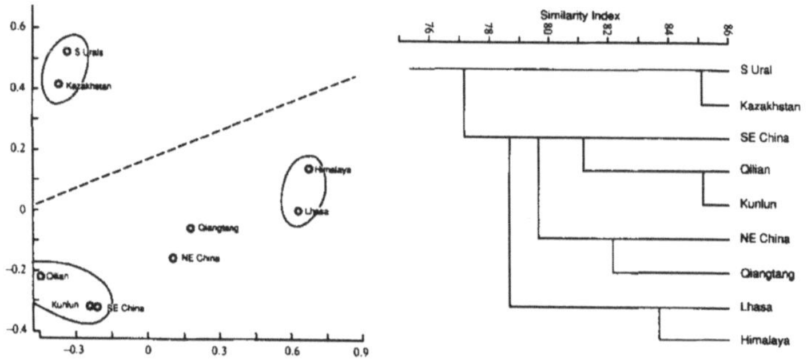

Figure 5.9 Analyses of Visean rugose coral genera, excluding all simple solitary forms. Left: first two principal axes of a PCORDS analysis; right: dendrogram from an SLC analysis.

Source: Smith (1988). Reproduced by permission of the Royal Society.

development of glacio-marine sediments and by the absence of both complex rugose corals and fusulinids. Interestingly, the major faunal break between 'Himalayan' and 'Cathaysian' is probably the product of a latitudinal/climatic gradient, rather than separation of terranes by an oceanic hiatus, because it crosses the Qiangtang terrane. Cathaysian region occupied a broadly equatorial latitude in the late Early Permian, with the highest diversity possibly in the South China Block and Qiang-tang terrane. Although NE China shares coral and fusulinid genera with these two blocks, it occupied a marginal position of this climatic belt. In the early Late Permian moderately high-diversity coral faunas are known from a number of areas. Highest-diversity faunas of rugose corals are recorded from Lhasa terrane and SE China. The most sign-ificant feature is the marked disjunction between broadly tropical biotas in the Lhasa, Qiangtang, Kunlun, North and South China blocks as compared to the higher latitude biotas of the Himalaya terrane to the south and NE Siberia to the north. The marked difference between the Lhasa and Himalaya faunas indicates latitudinal differences between these two areas.

On the basis of results obtained from the different numerical analyses, Smith was able to demonstrate the following (Fig. 5.10).

Box 5.2 continued

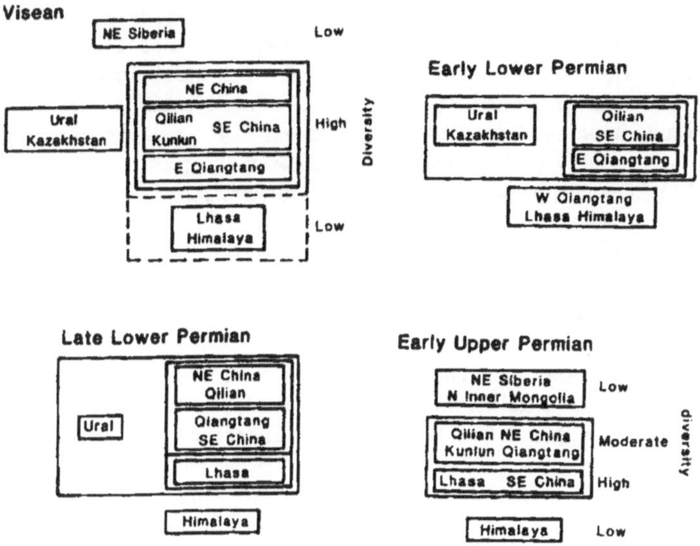

Figure 5.10 Diagrams showing the changing pattern of relationship for the biotas of regions within central East Asia deduced from the different numerical analyses performed.

Source: Smith (1988). Reproduced with permission of the Royal Society.

1 During the Carboniferous and early Permian the Cathaysian block formed one cohesive biotic region lying tropically or subtropically.
2 The southern boundary to the Cathaysian region does not coincide with a single suture zone through time, nor is it sharply defined. The apparent faunal impoverishment southwards across the Tibetan Plateau implies global climatic control (possible effects of the Carboniferous–Permian glaciation) instead of a geographic barrier.
3 The Lhasa and Himalaya terranes shared a similar fauna until the mid-Permian, when a marked faunal disjunction developed coincident with the Zangbo suture.

Box 5.2 continued

Subsequent palaeogeographic reconstructions based on geological and geophysical data (Ricou, 1994) confirmed Smith's results. The Late Permian departure of the 'Lhasa' terrane from Gondwana provoked the opening of the 'Neotethys'. Smith's results have also been taken into account in the palaeogeographic maps recently realized by C. R. Scotese in the framework of the PALEOMAP project. These maps are available in the website www.scotese.com.

The problem of the Ordovician Tornquist Sea

It is well known that Ordovician fossils of Scotland are more similar to coeval fossils from Canada and the USA, whereas faunas of Southern England and Wales show stronger similarity with those collected in Eastern Newfoundland and Nova Scotia. Fossils from the last two localities, in turn, are different from those of eastern Canada. Furthermore, a different Ordovician fauna has been recognized in Scandinavia and Russia. The Caledonian chain runs between these localities and represents the suture of an ancient ocean, called the Iapetus ocean (Wilson, 1966): this separated an equatorial North American continent from Gondwana and Baltica, which were located at southerly high latitudes. The trilobite distribution recognized by Whittington and Hughes (1972) through multivariate analysis of dissimilarities has confirmed the faunal affinities outlined above. As a result, the palaeogeographic positions of Southern England, Wales and Eastern Newfoundland were considered to have been close to Gondwana and Baltica. Cocks and Fortey (1982) refined the Ordovician and Silurian palaeogeographical reconstructions on the basis of palaeobiogeographical constraints. They inferred the existence of another ocean, the Tornquist Sea, which separated Baltica from a block called Avalonia. This block, whose palaeogeographical position has been located on the northern Gondwana margin, corresponds to Nova Scotia, Eastern Newfoundland and Southern England terranes. Cocks and Fortey importantly stated that the Tornquist Sea disappeared in the later Ordovician and joined Avalonia to Baltica in a single block, separated from Gondwana by the southerly Rheic ocean. Faunal exchange was suggested across Iapetus by Early Caradoc on the

Box 5.2 continued

basis of trilobite (Whittington and Hughes, 1972; Cocks and Fortey, 1982) and brachiopod (Williams, 1973) distributions which indicate increased affinities between Avalonia–Baltica and Laurentia. Paris and Robardet (1990) proposed a different reconstruction. Avalonia was considered to be part of Baltica since the Early Ordovician and probably since the Cambrian, thus rejecting the existence of the Tornquist Sea. The Rheic ocean, whose opening was inferred by Cocks and Fortey (see also McKerrow and Cocks, 1986) to have began during the Caradoc, was envisaged to separate Gondwana from Avalonia. According to Paris and Robardet, in the Early Ordovician the reduced width of the Rheic ocean would not prevent the faunal exchange, thus explaining the palaeobiogeographical affinities stressed by Cocks and Fortey. Figure 5.11 shows the two schemes. It is worth noting that from Silurian onwards the reconstructions of Cocks and Fortey and Paris and Robardet are substantially similar. Thus, the problem is: did Avalonia make part of Baltica or were these regions separated by a Tornquist Sea?

Fortey and Mellish (1992) analysed the distributional patterns of different Early Ordovician organisms (trilobites, ostracods, brachiopods, graptolites and acritarchs) using PAE. The authors processed the data for a single time-plane using the DOLLO option available in PAUP version 3.0 (Swofford, 1990). The cladograms produced for each taxon (Fig. 5.12) were compared with trees generated by single linkage cluster analysis. The patterns derived from these analyses clearly showed a threefold palaeogeographical division, of trilobites and ostracods in particular and of brachiopods and graptoloids to a lesser extent. The results of bootstrapping on trilobite trees support the threefold biogeographical division. The low CI for acritarchs suggested poor biogeographical discrimination by these planktonic organisms, which were less sensitive to the biogeographical effects of a marine barrier than the benthic ones. These three distinct areas of endemism have been considered as indicative of three continental masses, Laurentia, Baltica and Gondwana, separated by two oceanic hiatuses, the Iapetus ocean and the Tornquist Sea (Fig. 5.11). Fossil faunas from localities of Avalonia cluster with Gondwanan faunas. Thus, Fortey and Mellish rejected the reconstruction proposed by Paris and Robardet (1990). On the basis of palaeomagnetic data, Torsvik

Box 5.2 continued

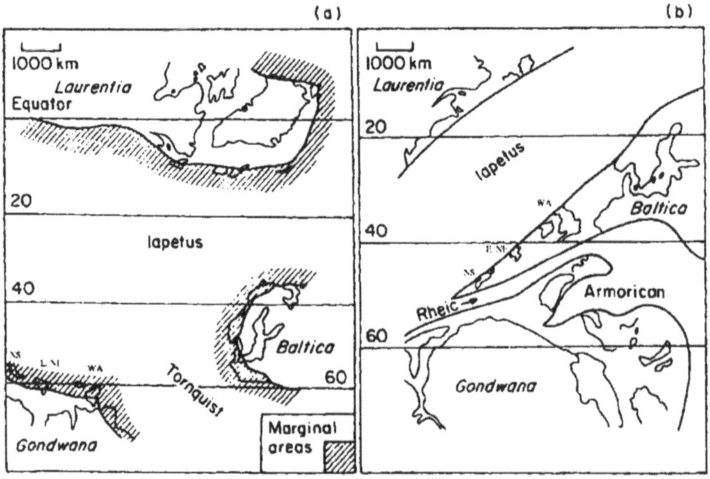

Figure 5.11 Palaeogeographical reconstructions of Early Ordovician time. Left: Cocks and Fortey's (1982) scheme showing Avalonia on the northern Gondwana margin and the Tornquist Sea. Note that more recent versions based on palaeomagnetism additionally rotate Baltica at this time (see Torsvik and Trench, 1991 and Torsvik *et al.*, 1995); right: Paris and Robardet's (1990) scheme without the Tornquist Sea and with Avalonia considered part of Baltica. NS, Nova Scotia; E. NF, Eastern Newfoundland; WA, Wales.

Source: Fortey and Mellish (1992). Reproduced with permission of the authors, Blackwell Science, Elsevier Science and the Geological Society of London.

et al. (1995) support the hypothesis of a Tornquist Sea in the Arenig with Avalonia located close to Gondwana. This reconstruction is substantially congruent with Fortey and Mellish's patterns and has been integrated in the recent maps of the PALEOMAP project. Recent acritarch data also confirm the Gondwanan position of Avalonia during the Tremadoc–Arenig transition (Servais and Mette, 2000).

Box 5.2 continued

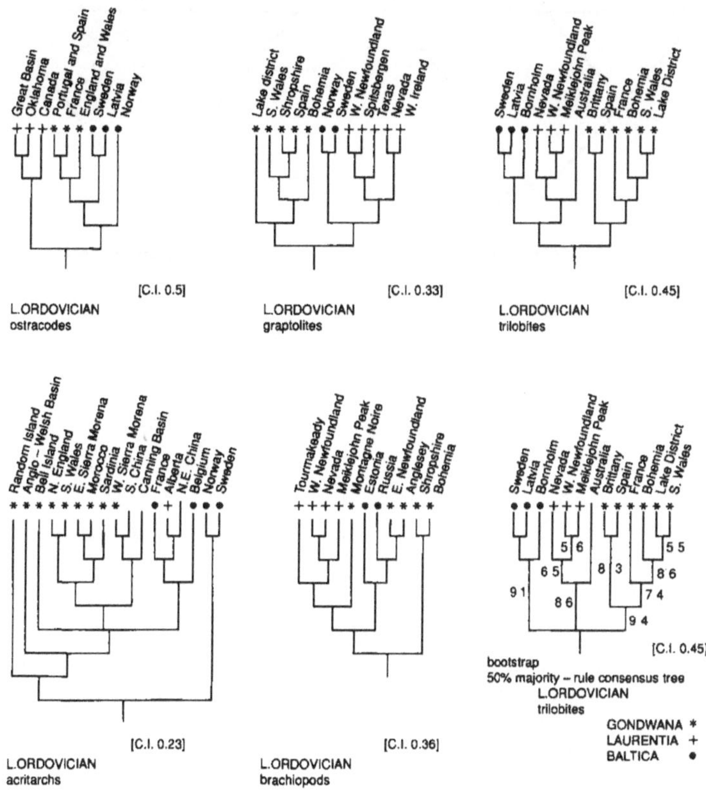

Figure 5.12 Analyses of hierarchical similarity of different Early Ordovician faunas using PAE. CI, consistency index.

Source: Fortey and Mellish (1992) and references therein. Modified by permission of Blackwell Science.

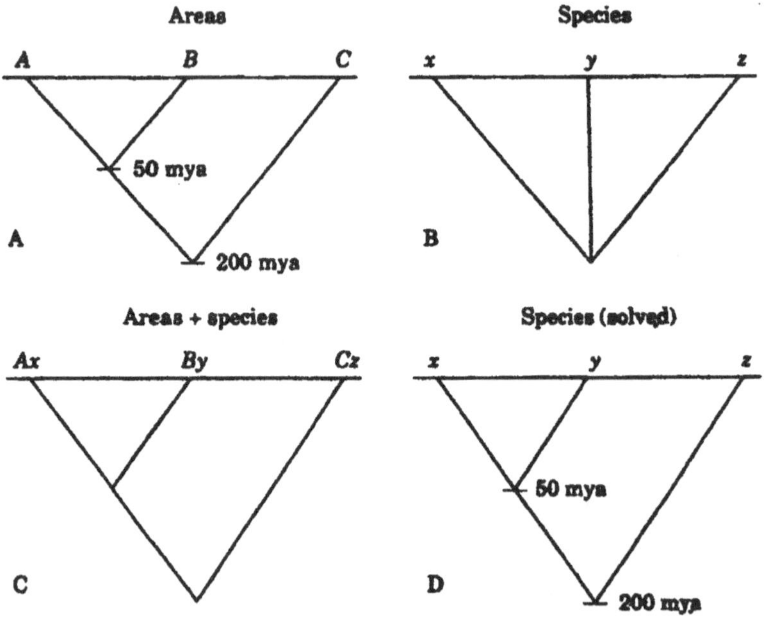

Figure 5.13 The technique of retrovicariance biogeography. Geological evidence indicates the time of separation of areas A, B and C (A). The phylogenetic relationship of taxa *x*, *y* and *z* is unknown (B) but as their geographic distribution is plotted in the area cladogram a phylogenetic hypothesis results (C). When the geologic time of area separations is known, the time of cladogenesis can be suggested (D).

Source: Monge-Najera (1995). Reproduced with permission of the author and Academic Press.

5.4 Retrovicariance biogeography

This method uses area cladograms, produced from geological data, as the basis for constructing biogeographic, systematic and palaeontological hypotheses, which can be tested against independent non-biogeographical evidence (Fig. 5.13). The requirements of retrovicariance are the same as those of vicariance, i.e. evolution by cladogenesis, monophyletic taxa and non-composite geographic areas.

This method has been applied to reconstruct the phylogenies of some taxonomically problematic organisms, for example velvet worms of the phylum Onychophora (Monge-Najera, 1995, 1996), but so far I am not aware of applications to fossils of marine invertebrates.

6 Larval biogeography

The spatial range of planktonic organisms is mostly controlled by abiotic factors (climate, currents, nutrients), whereas the biogeographic distribution of benthic organisms depends on biological characteristics that govern the process of dispersion. The early life-cycle of numerous invertebrates is characterized by a larval stage preceding the metamorphosis. During the larval cycle the organisms can be dispersed by currents at different distances from the place of birth.

In palaeobiogeography the larval transport is inferred through comparison with modern representatives of invertebrate groups. In some cases, particularly for gastropods and bivalves, the type and the possible duration of the larval stage can be reliably inferred from fossils on the basis of observable and measurable morphological characters of the embryonic shell.

The duration of the larval stage defines the dispersion potential of benthic organisms. On the other hand, physiological factors that are typical of each species constrain the growth and the reproductive history. Larval transport also depends on currents. Vectorial characteristics of the palaeocurrents are a matter of assumption or speculation to explain palaeobiogeographical patterns.

Larval development is considered a major tool for biogeographic interpretations. Thus, it has been intensively studied on extant organisms and the most important results are reported here, focusing on the most relevant biogeographical aspects. Criteria have been defined to recognize the larval types on the fossil shells, particularly the larval shells of gastropods and bivalves.

6.1 Types of larval development

Different terminologies can be found in the literature for marine invertebrate larvae (Table 6.1). In this book we refer to the classification

Table 6.1 Terminologies for marine invertebrate larvae (from Jablonski and Lutz, 1983, redrawn with permission).

	Thorson (1950) Ockelmann (1965)	*Scheltema (1978) Crisp (1976)*	*Shuto (1974) Strathmann (1978)*	*Jablonski and Lutz (1980, 1983))*
Free-swimming, feeding larvae	Planktotrophic		Planktotrophic	Planktotrophic
Free-swimming, non-feeding larvae	Lecithotrophic	Pelagic		
Offspring emerge as benthic juveniles	Direct development	Non-pelagic	Lecithotrophic	Nonplanktotrophic

proposed by Jablonski and Lutz (1983). Accordingly, the basic larval types are two: planktotrophic and nonplanktotrophic.

6.1.1 Planktotrophic larvae

As the name indicates, these larvae feed on planktonic, smaller organisms or organic particles. Their mode of life is planktonic and their development time is mostly spent swimming freely in surface waters. This implies the development of particular structures functional to locomotion and feeding, all of which are lost or resorbed during metamorphosis.

The extremely high rate of mortality of the planktotrophic larvae, which has been extimated to exceed 99%, is counterbalanced by the high fecundity: in fact, enormous numbers of larvae are produced in planktotrophic species. Although parental expenditure per egg is minimal because of little yolk content, the total parental investment in reproduction is actually very high. The ability of planktotrophic larvae to subsist on planktonic food obviates the need for deposition of high quantities of yolk in the egg as energy source, and also allows a long duration of the larval stage of many planktotrophic species.

The duration of the larval stage is long, spanning from two to six weeks. Considerably longer larval stages characterize the **teleplanic** larvae, whose duration can reach six months or even more. The organisms characterized by planktotrophic larval development can be transported over the greatest distances. Accordingly, the biogeographical distribution of such organisms, teleplanic larvae in particular, is potentially very wide. Disjunct patterns of biogeographical distribution are frequent.

In shallow waters of the continental shelves, species are overwhelmingly planktotrophic in low latitudes. The advantages of the planktotrophic development are the high dispersion power, increased gene flow, increased colonizing ability, recruitment success and greater opportunity to select suitable habitats (Valentine and Jablonski, 1982). On the other hand planktrophy decrease in frequency with increasing latitude and increasing depth.

6.1.2 Nonplanktotrophic larvae

Many different types of larvae are included in this category. The most common are the **lecithotrophic** larvae, which feed on the yolk of the eggs from which they develop. Their mode of life can be either nonplanktonic or planktonic for some hours or days only. The rates of mortality are lower and reproductive effort per offspring is much higher than in planktotrophic species. Therefore, the number of eggs per parent is considerably lower. This type of larval development has been observed in several bivalve and gastropod species as well as in brachiopods.

So-called '**mixed development**' occurs when early developmental stages are encapsulated but free-swimming, pre-metamorphic larvae emerge at later stages. This 'mixed' development is important in some benthic marine groups, including gastropods and polychaetes.

In contrast, ovoviviparous species have lecithotrophic larvae that develop entirely within an egg mass or egg case. The metamorphosis takes place within the protection of the egg mass and embryos may pass through a lecithotrophic veliger stage or not. The term **direct development** should be applied in the latter case for species that hatch directly into benthic juveniles, lacking a free larval stage altogether. This type of development has been shown in certain opisthobranchs, most cephalopods, Polyplacophora and some echinoderms.

Brooded larvae are characteristic of ovoviviparous species and are retained within parental body space during the development, eventually emerging as metamorphosed juveniles. These species invest the greatest amount of energy in the protection of their offspring during early ontogeny. Some gastropod, bivalve, ophiuroid and brachiopod species are brooders.

Dispersion power of nonplanktotrophic species is considerably lower than that of planktotrophic ones. The nonplanktotrophic larvae have restricted biogeographic distributions relative to the more cosmopolitan planktotrophic ones. Their distributional areas are more restricted but more continuous, disjunctions being rare.

In very high latitudes and in the deep sea, numerous species are nonplanktotrophic that either brood or are direct developers. Correlated with the increase of nonplanktotrophs in increasingly deeper waters is the general nonplanktotrophic strategy of small-bodied species.

6.2 Larval development in some groups of marine invertebrates

The known characteristics of the larval development of some groups of marine invertebrates are summarized below on the basis the data published by Jablonsky and Lutz (1983, with references therein).

6.2.1 Scleractinians

Most scleractinian species have a very wide geographical distribution which is attributed to long-distance dispersion of planulae by oceanic currents. However, direct information on coral larvae is limited because of difficulties in observing larval dispersion in nature. According to Harrison and Wallace (1990), reproduction in corals is usually seasonal; breeding occurs during brief annual periods. Most corals spawn eggs and sperm for external fertilization and development. Some coral species have brooded planulae which are produced asexually. Coral planulae swim more slowly than the current velocities and are essentially passive larvae, thus the duration of the planktonic stage is a fundamental factor in dispersion and settlement.

Brooded coral planulae rapidly settle (a few hours to a few days) near parents or within the natal reef after being released at an advanced developmental stage. However, their limited dispersion cannot account for the observed geographical range (2000 km) of some species.

On the contrary, the entire larval development of many gamete-spawning coral species occurs externally, in the plankton, and may last four to six days before settlement and metamorphosis. Shorter (two to three days after fertilization) or exceptionally longer (10 weeks) settlement periods have been observed, but may be laboratory artefacts. In any case the few days of planktonic development indicate that the majority of externally developed planulae may be dispersed away from their natal reef.

Both neo- and palaeobiogeographic patterns fuel debate over the extent of dispersion at different geographical scales. On a relatively small scale planulae are believed to be retained on their natal reef or dispersed between reefs, which are therefore either primarily self-seeded or interconnected for larval recruits. Predation on plankton is an important factor in evaluating the relative contribution to recruitment from planulae retained on the natal reef versus those dispersed to other reefs. In fact predation is higher within reefs than in open water, so retained planulae may provide few larval recruits. On a wider scale, the debate is focused on biogeographic importance of larval dispersion over long distance (for example the origin of the eastern Pacific coral fauna).

6.2.2 Brachiopods

Due to the importance of this group in the fossil record, particularly in the Palaeozoic, it would be necessary to know the developmental characteristics of the extinct taxa.

6.2.2.1 Rhynchonelliformea and Craniiformea

These subphyla were formerly included in the 'Articulate' group. Larval shell is missing because calcification begins only after settlement. The extant species have a nonplanktotrophic larval stage lasting from a few hours to a few days; larval brooding occurs in many of these species. However, the possible existence of the planktotrophic type of larval development during the Palaeozoic is inferred on the basis of the vast palaeobiogeographic distribution of some Palaeozoic forms (Valentine and Jablonski, 1983a). These seem to show latitudinal gradients characterized by a distinct low-latitude maximum similar to those of modern planktotrophs rather than modern nonplanktotrophs, including living articulates, which display a reduced or absent equatorial diversity maximum. However, some Palaeozoic palaeogeographic reconstructions are still controversial, thus the hypothesis of a planktotrophic development in Palaeozoic Articulata must be treated with caution.

6.2.2.2 Linguliformea

In this group the planktotrophic type of larval development has been recognized. It is recorded by a characteristic larval shell, the protegulum, which is recognizable in fossil forms. Various hypotheses on duration of the planktonic larval stage, based on the structure of the protegulum shell, have been proposed but still need to be tested.

6.2.3 Bryozoans

These are almost exclusively nonplanktotrophic: for example, 95% of modern cheilostomes have nonplanktotrophic larvae. However, little attempt has been made to infer mode of development in fossil bryozoans, though the distinctive larva of the planktotrophic forms is smaller than the nonplanktotrophic one and therefore distinguishable by measuring the initial zooid (ancestrula). However, it has been observed that the large nonplanktotrophic larva of several species differentiates into small founding zooids almost simultaneously. The larval ecology could be inferred from adult morphology. Brood chambers are often calcified, as in the ooecia of cheilostomes and the gonozooids of cyclostomes, thus making the ovoviviparous habit recognizable in some fossil lineages. But can the absence of skeletonized brood chambers be reliably taken as evidence for a planktotrophic larval stage? No firm conclusions can be currently stated.

6.2.4 *Bivalves*

The shell formed before the metamorphosis is called **prodissoconch**. The prodissoconch I and II are distinguished to indicate the different processes of growth of the larval shell. Prodissoconch I is relatively small (60–200 μm) and indicates the planktotrophic development. Among the nonplanktotrophics the species having lecithotrophic larval type, with a pelagic stage, possess relatively large prodissoconchs II (135–230 μm) whereas those characterized by the absence of a pelagic stage have the greatest prodissoconchs II (230–500 μm). Lecithotrophy, which corresponds to low dispersion potential, is the most common mode of development among deep-sea bivalves although maintenance of the gene flow and the recruitment over wide areas is suggested by the wide distribution of numerous deep-sea species.

6.2.5 *Gastropods*

The larval shell is called **protoconch**. Similarly to bivalves, protoconches I and II are distinguished. Protoconch I is relatively small and indicates a planktotrophic larval stage. Species characterized by lecithotrophic larvae with a pelagic stage have relatively large protoconches II. Species having no planktonic stage bear the largest protoconches II. Planktotrophic development has been observed to increase with depth in some prosobranch species and individuals. According to Bouchet and Taviani (1992), 40% of bathyal and abyssal gastropods are characterized by planktotrophic larval development. Their veliger larvae migrate from deep to surface layers during the ontogeny and feed on phytoplankton in epipelagic waters before returning to deep waters for metamorphosis.

6.2.6 Cephalopods

These molluscs are mostly direct developers: the hatched individuals do not undergo metamorphosis. Dispersion occurs through different processes: cephalopod eggs are contained in gelatinous egg masses that may remain fixed on the bottom or may drift with surface currents settling to the bottom only afterwards; passive transport of immature individuals has been observed too.

According to Landman *et al.* (1996), an important group of fossil cephalopods, the ammonites, probably underwent similar processes of dispersion. These authors proposed that ammonites' embryonic shell (called ammonitella) was probably neutrally buoyant and, consequently, they could have lived in the water column rather than on the bottom. In the plankton newly hatched ammonoids may have been active swimmers or, more probably, passive vertical migrators, drifting with surface currents. This planktonic stage may have lasted several weeks or months, depending on the size of the ammonitella, its rate of growth and the mode of life of the adult. A planktonic mode of life at hatching is consistent with the wide palaeobiogeographic distributions of numerous ammonite genera despite the fact that the adults are presumed to have been poor swimmers.

Compared to the extant *Nautilus*, the embryonic shells of ammonoids were ten times smaller. On the other hand, ammonoids produced numerous eggs, were probably semelparous (reproducing once and then dying) and were characterized by high mortality at hatching as dense concentrations of ammonitellas in the fossil record suggest. Compared to Ammonitina, Phylloceratina and Lytoceratina have larger embryonic shells and are characterized by wider palaeobiogeographic distributions correlated with their longer geological life (see section 6.5.1).

6.2.7 Echinoderms

Larvae of some echinoderms, particularly Echinoidea, have a calcified skeleton, but fossil larvae of echinoderms have not yet been recognized. Although it is possible to infer the developmental history of some groups through careful examination of the adult morphology, the larval skeleton can partly form the nucleus for some adult plates or is partly resorbed during metamorphosis.

Planktotrophic larval development has been inferred from the diameter of the eggs of some deep-sea ophiuroid and seastar species but Asteroidea lack calcification of the early growth stages. The extant Crinoids are exclusively nonplanktotrophic, though some Palaeozoic forms could have been planktotrophics (Strathmann, 1978). Planktotrophic and nonplanktotrophic development in echinoids can be inferred in three different ways (Jeffery, 2001). Mueller (1970) suggested that the genital pore diameter reflects egg size, which is larger in nonplanktotrophs. The orientations of crystallographic axes of apical disk plates are used to deduce the developmental mode since the calcite skeletal rods of planktonic larvae determine the orientation of the calcite in the apical system of the adult echinoid. Brooding is accompanied by a skeletal modification of the parent, the marsupium, a sort of depression in the adult test which serves for brooding of juveniles.

6.3 Rates of dispersion

The potential amplitude of the dispersion depends on the larval development, planktotrophic versus nonplanktotrophic, but also on values of current velocities.

Some very rough reckonings can help us to have an idea of the extent of the dispersion. The westerly equatorial surface current of the Pacific Ocean can flow at 25–75 cm/s: a larva can travel for 21 to 65 km per day. Taking an average duration of six weeks for a species having a planktotrophic (non-teleplanic) larval development, in a single generation the larvae could travel over 882 to 2370 km.

Deep currents must also be taken into account. For example, in the Pacific Ocean at depths between 50 and 300 m, the Cromwell equatorial countercurrent flows easterly at 100–150 cm/s. Accordingly, a larva can travel for 86 to 130 km per day and a planktotrophic larva can travel over 3612 to 5460 km in six weeks. These values are, potentially, higher for teleplanic larvae.

The real course of the transport is certainly not as simple as we have supposed to reckon theoretical distances. Numerous factors, physical and ecophysiological, play an important role. Nevertheless, the values above are likely because the biogeographical disjunct distributions of modern molluscs imply the crossing of wide barriers (e.g. the East Pacific Barrier) without any oceanic islands that can be used as stepping stones.

6.4 Palaeocurrents

After examining the role of the larval transport and its potential for biogeography, two questions arise. 1 Can the existence and the main direction of palaeocurrents be deduced from biogeographic distributional patterns? 2 Can the palaeocurrents modelled by independent methods (palaeogeography, numerical simulations, etc.) explain the patterns? The literature provides numerous positive answers to both questions, and interesting arguments of biogeographic significance can be made for 'pure' and/or 'applied' (both evolution and palaeogeographic reconstructions) palaeobiogeographic purposes. Some examples are summarized below.

The affinities of foraminifers which colonized the Sergipe Basin (North-East Brazil) in Late Aptian–Albian times with North Atlantic–Western Tethyan assemblages are consistent with surface water palaeocirculation patterns (Koutsoukos, 1992). According to Chirat (2000), taphonomic evidence of post-mortem drift of empty shells of the nautiloid *Aturia* fits remarkably with the modelled palaeocurrents for the Palaeocene–Miocene interval. This explains the cosmopolitan pseudodistribution of this genus across barriers which presumably it could not cross. The most important Eocene to Pliocene distributional changes (range expansions and contractions) of pteropod faunas as well as faunal exchanges between Caribbean, Mediterranean and Paratethys since the Oligocene have been considered consistent with modern circulation patterns by Bernasconi and Robba (1982).

Néraudeau and Mathey (2000) have tested palaeobiogeographic patterns of South Atlantic Cretaceous echinoids against modelled patterns of oceanic surface circulation. For example, the striking similarity between Late Albian–Early Cenomanian faunas of Brazil and Angola, which include many common species, is consistent with the modelled anti-clockwise pattern of currents that probably flowed between South America and Africa (still not far apart at that time). On the other hand, the authors also considered that in the Late Cenomanian typical Mediterranean echinoids reached the Gulf of Guinea and Brazil through the trans-Saharan seaway more than by dispersion by currents (Fig. 6.1). Information on larval development of these echinoids would benefit the testing of biogeographic data with the modelled circulation patterns.

Larger benthic foraminifers constitute a key geoecological assemblage because these organisms prefer tropical and subtropical habitats and are common and highly diverse at low latitudes. During the Late Cretaceous these organisms were commonly restricted to the American side of the Atlantic at low latitudes, but their occurrence and relative abundance up to high latitudes of Europe (Holland, southernmost Sweden) led Fourcade and Michaud (1987) to infer the existence of a palaeo-Gulfstream which transported foraminiferal embryos. However, rudists, another element of the shallow platform biota commonly associated with larger benthic foraminifers, also occur in these Northern

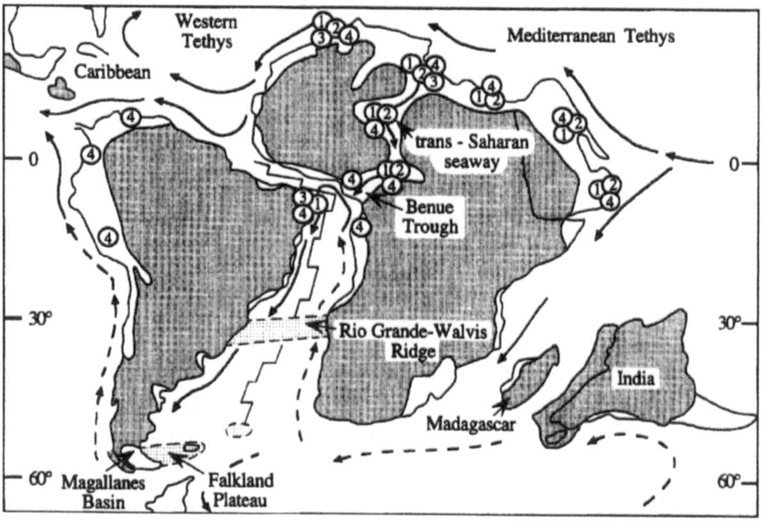

Figure 6.1 Distribution of Late Cenomanian echinoids of supposed Mediterranean origin (ringed numbers) with the main surface currents.

Source: Néraudeau and Mathey (2000) and references therein. Reproduced with permission of Elsevier Science.

European areas. This provides a biogeographic testing for the hypothesis of Fourcade and Michaud. The high level of endemism for Caribbean and Mediterranean rudists in the Campanian–Maastrichtian interval (Coates, 1973; Philip, 1982) would contradict this hypothesis, since most of the larger benthic foraminifer genera (*Orbitoides, Lepidorbitoides, Omphalocyclus*) common to American low-latitude Atlantic coasts and Northern Europe also existed in Southern Europe, Northern African and Mediterranean Tethys. Conversely, the hypothesis of a palaeo-Gulfstream could be supported if one considers the endemism of most Late Cretaceous rudist genera as reflecting non-planktotrophic larval devclopment. Unfortunately, no information on larval development of these peculiar bivalves is available.

The sudden appearance of larger benthic foraminifers in Western and Southern Australian coasts during the late Middle Eocene is suggestive of west-to-east warm currents. This record has been interpreted by McGowran *et al.* (1997) as the beginning, earlier than previously thought (Miocene), of the Leeuwin Current which flows today from Western to South-Western Australia, causing there the absence of a cool-temperate province.

Transoceanic communications between Eastern Panthalassa and Palaeotethys during Late Carboniferous to Permian times have been inferred by Lethiers and Crasquin-Soleau (1995) by means of benthic shallow-water ostracod faunas. On the basis of values of the Johnson's index, the most significant faunal similarities have been shown to exist between the western margin of Palaeotethys (with the exception of Israel) and North America. According to Lethiers and Crasquin-Soleau, the inferred communications could be explained by east-to-west surface palaeocurrents (North Equatorial Current and South Equatorial Current) which carried ostracods on biological rafts (floating algae already existing in the Late Palaeozoic). Furthermore, intraoceanic staging posts (suspect terranes, guyots) have been invoked to explain the high number of transpalaeopacific ostracod species. The same scenarios have been considered by Vachard and Fourcade (1997) to explain fusulinid palaeobiogeographic patterns in the Permian. The faunal affinities between Oman (Palaeotethys, high latitudes) and North America (Eastern Panthalassa, low latitudes), in spite of the different palaeolatitudes deduced by the palaeogeographic reconstruction of Baud *et al.* (1993), are explained by anti-clockwise gyral circulation enhanced by southern deflection of the warm South Equatorial Current (modelled by Ziegler *et al.*, 1981) by the southern pangaean continental mass. The low similarity between North America and Bolivia (high-latitude Eastern Panthalassa) is explained by the influence of cold currents originating from polar South-Gondwana. Similarly, cold currents are inferred to explain the lack of affinities between Australia and North America, in contrast with the circulation model of Ziegler *et al.* (1981). Interestingly, all the species shared between these distant areas belong to genera having benthic larvae. Ostracods are not considered to be able to migrate over long distances because they are probably nonplanktotrophic (Babinot and Colin, 1992) and their wide distribution is probably due to the drifting of algal masses that they may live in (Whatley, 1988).

6.4.1 Risks

Often, the results of a biogeographic study roughly coincide with independently modelled palaeocurrents which are continuously subject to revision in the light of ongoing research. Thus, the biogeographical conclusions may fit the reconstructions in an uncritical way. There is also a strong risk of circularity because in numerous studies the oceanic currents are reconstructed on the basis of the biogeographic distribution of particular taxa, both to explain biogeographical pattern

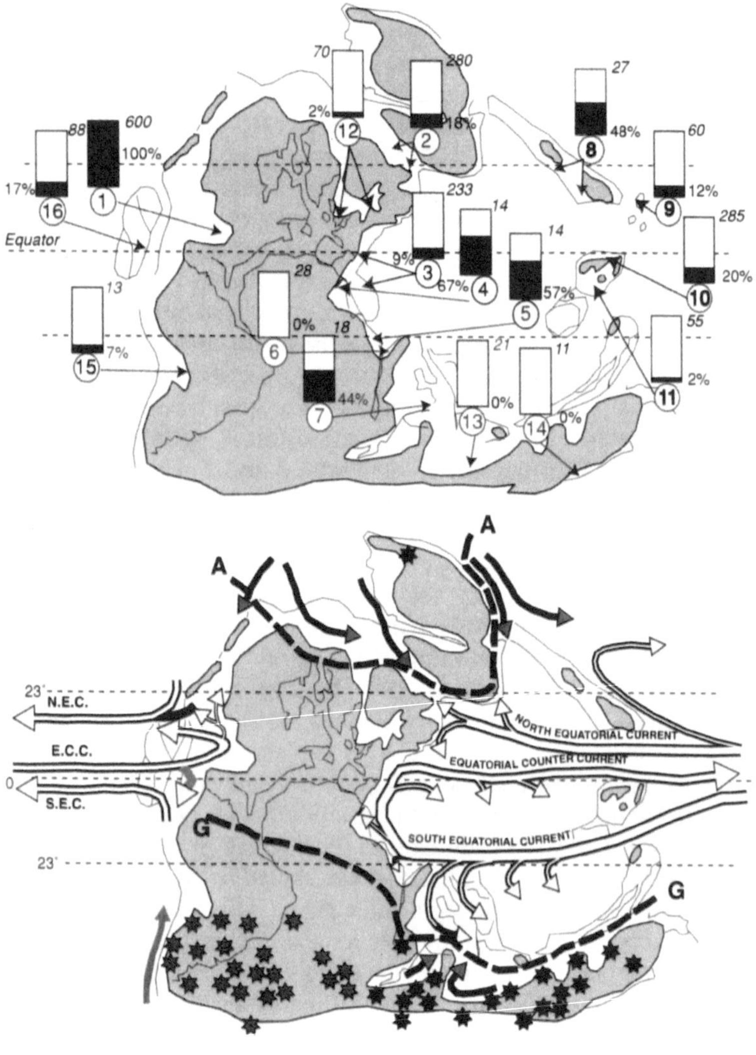

Figure 6.2 Rates of common latest Carboniferous and Permian ostracod species (dark part of rectangles) of different localities (encircled numbers) relative to USA (taken as 100%) are indicated above; the numbers in the upper right corner of rectangles correspond to the species number at each locality. Modelled palaeocurrents for Permian are represented below: white arrows indicate warm currents whereas black arrows correspond to cold surface currents. Thick lines represent schematic Permian coastlines and shaded pattern indicates lands. Stars indicate traces of glaciation. A, southern boundary of the Angaran Realm; G, northern boundary of Gondwanan Realm.

Source: Lethiers and Crasquin (1995). Reproduced with permission of Geobios.

itself and to infer the existence and the direction of a palaeocurrent (which is believed to be the only factor controlling the dispersion).

Palaeocurrents are thought of in an oversimplified way and their biogeographic effects are cumulative relative to the vectorial characteristics. Currents are in fact extremely complex systems which also include water masses flowing in opposite directions; current directions may change because of seasonal changes of general circulation. On the other hand, palaeocurrents are 'time-averaged': for example, a hypothetical future sedimentary record of present-day westward equatorial current in the Pacific Ocean will show the effects of the main direction even though its flow is periodically inverted in coincidence with 'El Niño' events. Because of the physiologic requirements of the organisms, in biogeography the quality of the vector (warm or cold currents) is dramatically important (Bhaud, 1989).

An evolutionary value is often recognized to palaeocurrents inferred from palaeobiogeographic, long-term, patterns. For example: a given taxon of a given age is recorded in a certain area and a sister taxon occurs in another area in a subsequent time. This kind of pattern is generally used to infer both a biogeographic-oriented evolution of a group and the existence of a palaeocurrent with its vectorial properties. The interval of time that was necessary to produce such a kind of palaeobiogeographical pattern is generally unknown but certainly a long-time one, whereas the current circulation could produce 'instant' effects at short-term intervals of time. Therefore, biogeographic patterns similar to the one described above appear to be the cumulative result of numerous short-term processes, which may be also the result of different histories of current systems. Roux (1982) suggested more care in invoking the effects of currents to explain with geometric and phyletic criteria the relative directions of distributional changes. This is the case of the occurrence of Campanian–Maastrichtian rudists, larger foraminifers and associated shallow-water tropical organisms from Caribbean to Eastern Tethys (Schlanger *et al.*, 1981; Skelton and Wright, 1987; Skelton, 1988). This has been interpreted as a range expansion from an inferred Caribbean centre of origin favoured by intraoceanic shallow-water staging posts. As far as rudists are concerned, a different scenario has been proposed by Philip and Platel (1994, 1998a, 1998b) on the basis of both the evolutionary history of the hippuritid genus *Torreites*, later contested by Morris and Skelton (1995), and the radiolitid genera *Tampsia* (endemic to Mexico and Cuba) and *Monopilarites* (so far only reported from Oman). A more complex relation between the observed biogeographic distribution of these rudists and the modelled palaeocurrents is envisaged by Philip and Platel, who suggest to replace the east-to-west equatorial current

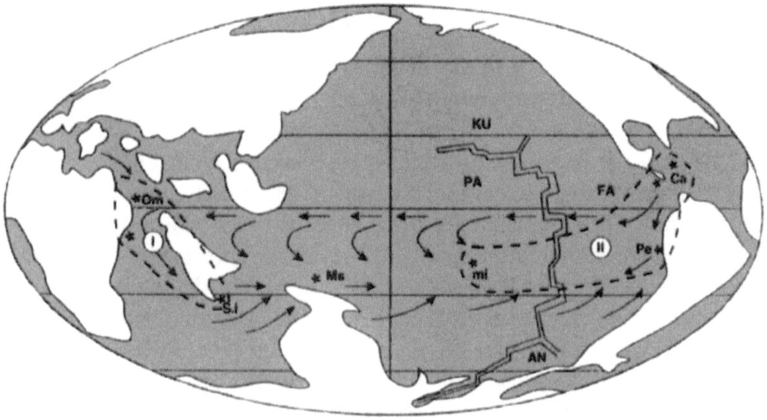

Figure 6.3 Rudist palaeobiogeographic distribution during the Maastrichtian and oceanic surface currents. I, Arabo-African and Indian Province; II, Caribbean and East-Pacific Province. Asterisks correspond to rudist occurrences. Abbreviations: Om, Oman; S.i, South-East India; Ms, Misol; mi, Marshall Islands; Pe, Peru; Ca, Caribbean; PA, Pacific Plate; FA, Farallon; KU, Kula; AN, Antarctic.

Source: Philip (1998) with references therein. Reproduced with permission of Société Géologique de France.

(Fig. 6.3) by a gyral circulation in the southern hemisphere (Barron and Peterson, 1989; Philip, 1998). Whatever phylogenetic reconstruction is held, the total lack of information about the characteristics of the larval development of rudists hampers the assessment of the role of oceanic currents in determining the observed biogeographic patterns.

6.5 Larval development and evolutionary implications

6.5.1 Planktotrophic species have longer geological life

Both the geological life and the rates of speciation are correlated with the biogeographic dispersional characteristics of larvae. The lower rates of speciation have been recognized within the planktotrophic organisms because the long duration of the larval stage allows the preservation of the genetic flow between disjunct populations of benthic adults, thus avoiding both genetic divergence and eventual speciation processes. The probability of speciation seems to decrease, therefore, with increasing geographic range (Flessa and Thomas, 1985). Conversely, the nonplanktotrophic species tend to have higher speciation rates. Their potential of dispersion is lower and they generally develop local populations that may remain isolated after the original colonization or the splitting (vicariant event) from a parent population. Therefore the genetic flow can be easily interrupted, thus promoting speciation processes. If species characterized by a long larval stage are less likely to become isolated than species with a short larval stage, one can argue that as the distributional area increases the chances of subdivision of that area by vicariant events also increase. This could increase the probability of speciation with increasing geographic range. However, it is not likely, and it is not demonstrated, that vicariant events affect all species equally. Events that caused the geographic isolation of nonplanktotrophic species may not affect planktotrophic species, which are characterized by longer dispersional stages and wider ranges.

The wide distributional areas of species characterized by planktotrophic larval development imply that local catastrophic events can eliminate a single population but not the whole species over its total distributional area. Therefore the persisting populations will provide new larvae, i.e. new potential adults, to the area where the original population of the same species has been decimated by a local extinction. Eurytopy may be significant in species with planktotrophic development because environmental conditions may vary from parental area to settlement site. All these elements produce lineages characterized by low extinction rates and geologically long-lived species. On the contrary, the nonplanktotrophic species are characterized by restricted and more continuous distributional areas (Jablonski, 1986b, 1986c). Therefore the species characterized by the latter type of larval development are vulnerable to local catastrophic events, which can lead to extinction, and result in shorter geological life (Fig. 6.4).

Nevertheless, Valentine and Jablonski (1986) have shown that gastropod genera with planktotrophic larval development survived the end-Cretaceous extinction equally well as nonplanktotrophs. However, in mass extinction events the sorting of the survivors seems to depend more on characters of the biogeographic distribution (nontropical distribution, wide biogeographic range) than on a taxon's individual properties (Jablonski *et al.*, 1985).

The considerations outlined above are valid for benthic organisms. High dispersion coupled with large population sizes connects distant populations, reduces intraspecific variability and slows speciation rates. However, different evolutionary dynamics have been suggested for pelagic taxa by Norris (2000), who noted that high dispersal and large population size have not impeded the high rates of speciation observed in pelagic taxa. Norris explained this paradox envisaging parapatric models of speciation such as depth parapatry and seasonal parapatry,

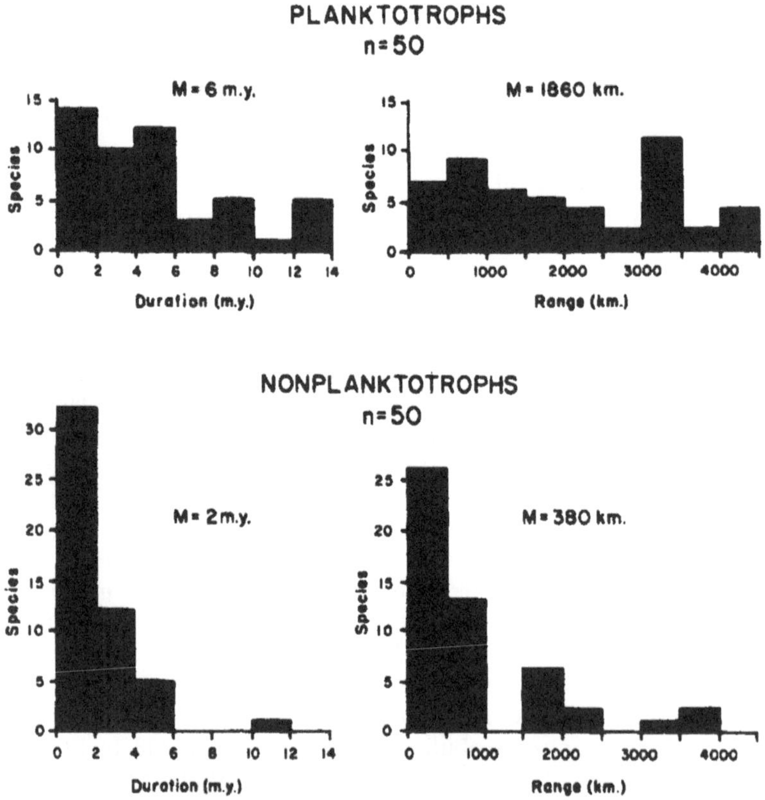

Figure 6.4 Geological durations and geographic ranges are greater for planktotrophic than for nonplanktotrophic species, as shown by a study of Late Cretaceous gastropods of the Gulf and Atlantic coastal plain, North America. M = median values; m.y. = million years.

Source: Jablonski (1986c). Reproduced with permission of the Rosentiel School.

which involve divergence between populations near the limits of the ancestral species' range, as well as sympatric models of speciation.

6.5.2 Metapopulations

A patch is 'the area of space within which a local population lives' (Hanski and Gilpin, 1991). The patchy geographic distribution of most species may be related to a physically discontinuous distribution of their life-environments. Patches are connected by migration or chance processes of transport and recruitment which strongly depend on dispersion power defined by the critical dispersion distance (see Chapter 3). For example, the development of reefs depends on suitable physiographical, climatic and oceanographic conditions whose distribution is discontinuous. In other words, populations of species inhabiting the different patches may be interconnected, depending on the limits of larval transport of benthic species or the capacities of active movement of nectonic species. The species concerned do not have a disjunct distribution because their distributional area includes the patches. These species are best regarded as a set of local populations which may be interconnected by patterns of gene flow, extinction, and recolonization. This set of local populations, which can be considered a population of subpopulations interacting by means of the movements of individuals among populations of different patches, therefore constitutes a **metapopulation** (Jackson *et al.*, 1996). This concept plays an important role in modern ecology, particularly for conservation purposes (Margules and Pressey, 2000) and in connection with disturbance events (Whittaker, 1998). However, metapopulation models are important not only in palaeoecology and macroevolution, but also in palaeobiogeography. The integration of this model with the concepts of larval biogeography may provide new insights in reconstructing the evolution of some biogeographic provinces. This concept can be associated with that of the staging post (see Chapter 3): the latter represents the physical frame for the metapopulation, which represents the biotic component.

6.5.3 Dispersion power and gene flow: evolutionary implications

The relations between the maintenance of the gene flow and the characteristics of the larval development, particularly the dispersion power, have important evolutionary consequences.

In modern seas, observed patterns of endemism at different taxonomic levels show high provinciality created by topographic and climatic barriers. Current systems make extremely complicated the theoretical, simple relation between climate, latitude and faunal distributions. Therefore in palaeobiogeography the inference of climatic barriers is often speculative. On the other hand, topographic barriers within marine palaeoenvironments may be detected through the record of diagnostic sedimentary successions and fossils. If the transition from marine to terrestrial environments is an obvious barrier, the factors that limit the extent of shallow organisms are often unclear and not always related to the topography in a simple way: the presence of emerged lands or the increasing of the depth at the shelf break etc. The thermocline is considered the most important barrier separating shallow from deep-sea biotas (Valentine and Jablonski, 1982, 1983b).

The patterns of speciation are also complicated by the physiographic characteristics of the marine environments. As far as shallow biotas are concerned, Valentine and Jablonski (1982) have introduced a distinction of three types of geographic shelf 'provinces' or shelf types. ***Type I*** is common today (for example the modern Pacific coast of North America) along north–south, linear continental shelves which are essentially one-dimensional, hundreds to a thousand km long and only 50–60 km wide. The area between the shoreline and the thermocline is narrow. ***Type II*** corresponds to a two-dimensional shelf setting such as regions of gentle bottom slope or broad epicratonic seas, with a wide area between the shoreline and the thermocline. Patchy distributions occur in this type of shelf because gentle topographic variations may produce embayments or interrupt the marine environment with islands etc. Uncommon today, this type of shelf was common during some geological periods (e.g. Devonian, Cretaceous). ***Type III*** is typified by the modern Indo-Pacific area, where the habitat distribution is discontinuous, with islands and small continental margins separated by oceanic depths.

Nonplanktotrophs have low powers of dispersion which restrict the gene flow. In *Type I*, linear shelves, species partitioning into isolated populations enhance vicariance. On the contrary, planktotrophic lineages are characterized by species with high powers of dispersion that maintain the gene flow over great distances. Only strong ecological barriers and great geographic distances can lead to isolation and then to speciation. In *Type I*, linear shelves, allopatric speciation should be uncommon among species with planktotrophic larvae because of the high gene flow. The north–south linear character of the distributional patterns of modern shelf species introduces a latitudinal climatic gradient that affects the populations living at the end-points of the distributional area. Gene flow can be interrupted by tectonic and/or eustatic events. In *Type II*, two-dimensional shelves, patchy distributions are due to the geographic variability. In *Type III* settings, the dispersion and the maintenance of the gene flow among scattered areas is favoured by planktotrophic development. Valentine and Jablonski (1983b) considered founder events as the major speciation pattern. Both the great species richness observed in numerous extant Indo-Pacific clades and the high percentages of endemic forms suggest common founder speciations. Larvae can be occasionally transported beyond their normal distributional area: in this case colonization can be successful if the environmental parameters fit the physiologic tolerance of the species or non-reproductive 'pseudopopulations' (see the example in Box 6.1) may occur. Also, adaptive divergence in small allopatric populations might lead to speciation.

Box 6.1 Pseudopopulations: the example of the modern deep-sea benthic fauna of the Mediterranean

The problem of pseudopopulations in biogeography can be illustrated by the interesting study of Bouchet and Taviani (1992), which I summarize below, on the present deep-sea benthos of the Mediterranean. It is necessary to summarize here some general information on this peculiar fauna. It is characterized by a very low degree of endemism and a low diversity compared to the fauna of the north-eastern Atlantic. Two-thirds of the macrofauna species have been recorded in depths shallower than 200 m; 8.6% of the total fauna lives exclusively below 200 m, 2.5% below 500 m and 0.3% below 2000 m. Furthermore, this extremely low diversity is not compensated by high density or high biomass. These conditions are apparently more severe in the Eastern Mediterranean. This deep fauna mainly consists of eurybathic species; the degree of endemism decreases with increasing depth and endemic species are rare and endemic genera exceptional. Data on deep-sea Pliocene to (?) mid-Pleistocene molluscs (Raffi and Taviani, 1984) and Pleistocene bryozoans (Rosso and Di Geronimo, 1998) suggest that the modern Mediterranean deep-sea benthic fauna is impoverished. Faunal characteristics similar to those of the present one were established during a glacial episode (Sicilian) when, interestingly, faunal similarity at the specific level between Atlantic and Mediterranean increased. The onset of post-glacial hydrological conditions and temperature-related hydrological perturbations are probably the causes of the decline of the richer and more diverse deep-sea Pleistocene fauna. The present-day hydrological conditions of the Mediterranean are characterized in the western basin by a homothermic water column below 200 m (temperature of 13°C), while in the eastern basin the water column is homothermic (13.5–14°C) below 400 m. Salinity is higher in the eastern basin.

An obvious question arises now: why has the Mediterranean deep-sea benthos not been replenished yet by Atlantic species? The action of the physiographical barrier represented by the 280 m deep Gibraltar sill is the classical answer. However, according to Bouchet and Taviani (1992), the larval development and ecology as well as the hydrology of the Mediterranean bottom water must be taken into account. Most holoplanktonic species, which are able to reach the epipelagic layers through vertical migrations, can cross passively the sill of Gibraltar via the inflowing Atlantic surface current. The latter allows a larva to travel passively 20–115 km eastwards in 24 hours. The chances of survival of the different species depends on their eco-physiological requirements, for the hydrological conditions of the Mediterranean differ significantly

Box 6.1 continued

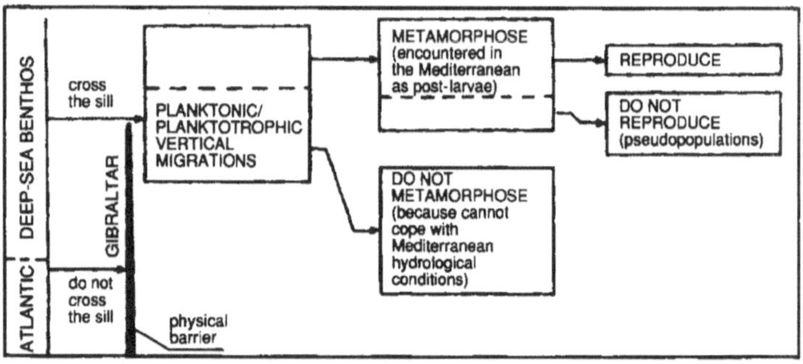

Figure 6.5 Successive filters encountered by Atlantic deep-sea benthic species when colonizing the Mediterranean.

Source: Bouchet and Taviani (1992). Reproduced with permission of Elsevier Science.

from those of the Atlantic (salinity is lower than the western Mediterranean basin and the temperature gradually decreases down to 2.5°C at 3000 m). Because the holoplanktonic fauna of the Mediterranean is an impoverished Atlantic fauna, the main biological barrier is not the sill of Gibraltar but, instead, the hydrological conditions of the Mediterranean waters.

The fate of larvae of deep-sea benthic species with a meroplanktonic larval stage can be paralleled to that of holoplanktonic species. Planktotrophic larvae of deep-sea benthic species capable of ontogenetic vertical migrations in the upper 200 m water layer are also passively transported from the Atlantic by the same inflowing current. They may therefore cross the sill of Gibraltar and eventually enter the Mediterranean. Thus, both holoplanktonic species and planktotrophic meroplanktonic larvae are not affected by this physiographical barrier; instead, the first barrier for these larval migrants is hydrological (Fig. 6.5) and actually corresponds to a filter because it allows the penetration of some species.

Meroplanktonic larvae migrate into deep, colder, waters for metamorphosis, but in the homothermic conditions of the Mediterranean they will not encounter waters colder than 13°C. This will not satisfy the physiological requirements of stenothermic species which, at best, occur in the Mediterranean as unmetamorphosed larvae. As explained in Chapter 1, the successful settlement does not depend simply on crossing a barrier and/or overcoming the critical period of metamorphosis.

Box 6.1 continued

In fact, the homothermic conditions of the Mediterranean may prevent reproduction of Atlantic species which, therefore, constitute pseudopopulations constantly maintained by a continuous dispersion. Therefore, only a small fraction of Atlantic deep stenotopic species may bypass the successive filters represented in Fig. 6.5. The pseudopopulation hypothesis of Bouchet and Taviani explains why the deep-sea benthic fauna of the eastern Mediterranean basin is poorer than that of the western basin: otherwise planktonic larvae of western reproductive populations could easily reach the eastern basin, crossing the sill of the Sicily channel. The hypothesis also explains the higher richness of the deep molluscan fauna during the glacial periods because the bottom of the Mediterranean was significantly colder, allowing stenothermal cold Atlantic deep-sea species to metamorphose. Consequently, neither tectonic lowering of the sill of Gibraltar nor current reversals are required to explain the occurrence of these species since the Sicilian.

Biogeographically speaking, the Mediterranean can be treated as an 'island' and the Atlantic as a 'mainland'; the equilibrium theory of island biogeography (MacArthur and Wilson, 1967) can then be applied. In this way the modern deep Mediterranean biota would represent an equilibrium between immigration (larval influx) and extinction (incapacity to settle or reproduce). The present interglacial hydrologic conditions began less than 15,000 years ago, a time-span probably too short to replenish the Mediterranean deep-sea by modern speciation events.

The hypothesis of Bouchet and Taviani is supported by empirical evidence. These authors primarily based their hypothesis on gastropods while believing that at least parts of their conclusions could be valid for many other deep-sea benthic animals. The interpretation of the modern deep bryozoan populations of Atlantic species in the Alboran Sea as relic populations, instead of pseudopopulations (Rosso and Di Geronimo, 1998), is probably justified by uncommon planktotrophic larval development in this group of organisms.

7 Palaeobiodiversity and palaeobiogeography

The relationships between surface area and biodiversity expressed in island biogeography or the present geographical distribution of biodiversity (strikingly shown in latitudinal diversity gradients peaking in the tropics and decreasing towards the poles) have been studied under the general heading of ecological biogeography. This is aimed at interpreting spatial distributions of assemblages relative to ecological parameters. Such an approach may, however, be too rigid because causal explanations of observed biotic distributions must take into account both ecological and time-related elements. Ecological analyses can be carried out at different spatial scales or levels of perception (see the extensive discussion in Rosen, 1988a), including the regional and continental scales that are most often considered in palaeobiogeography. In fact, it is extremely difficult to generalize diversity analyses at the community level at the scale of the overall diversity of an entire sedimentary basin or province.

Although a complete understanding of the various geological, palaeobiological and palaeoecological processes is required to account adequately for palaeobiogeographical patterns, fossil assemblages contain an ecological signature. This may be strongly biased by biostratinomic and taphonomic processes, but Hallam (1994) was right in suggesting that 'perhaps an admittedly coarse-grained account of diversity changes over several hundred million years may at least point the way to a more satisfying interpretation'. Biogeography may have a central role in explaining the relations between patterns observed at the community level with global diversity patterns.

This chapter will focus on those aspects of ecological biogeography relevant to palaeobiogeography, and in particular those dealing with the relationships between the compositions of palaeocommunities and their palaeoenvironmental settings, between the geographical distribution of biodiversity and species/area relationships, and the complex interactions between biodiversity and sea-level changes.

7.1 Biodiversity in palaeontology

Biodiversity can be expressed simplistically by the number of species. However, this approach fails to express the true dimensions of diversity because it neglects the relative abundance of species within the community. In palaeontology it is open to criticism because biostratinomic, taphonomic and diagenetic processes may delete or alter the original biological and ecological signature. The bias of these processes must be evaluated in any kind of palaeontological study (Brown, 1988).

Biologists usually treat diversity on the basis of a spectrum of quantitative indices that may apply to a variety of predictive situations with practical implications. Palaeontologists use quantitative treatments much less frequently because of the incomplete nature of the fossil record. Quantitative studies of the compositions of fossil assemblages are based on different techniques because taxonomic interpretations, particularly at specific and generic levels, are largely subjective.

7.1.1 Types of diversity

The term **richness** indicates the number of species forming a community, whereas **heterogeneity** indicates how these species are distributed within the community. The quantitative preponderance of either a very few or a single species over others is referred to as **dominance.**

The richness can be expressed by Margalef's index:

$$D = (S - 1)/\log_e N \qquad (1)$$

where S is the number of species and N the number of individuals. However, assessment of the diversity of a community also requires weighting of the constituent species by their relative abundances within the community. One of the most commonly used indices of diversity is that of Shannon-Weaver (Wiener):

$$H' = \sum_{i=1}^{s} (p_i)(\log_2 p_i) \qquad (2)$$

where p_i, the frequency of a species within the community, corresponds to n_i/N (n_i is the number of individuals of a species and N the total number of individuals of the sample). Any logarithmic base can be used. For the same number of species, H' increases with evenness and reaches its maximum value when all species of a sample are equally frequent. If only one species is present in the community, $H' = 0$. Etter (1999) noted that the Shannon-Weaver index is most sensitive to changes in rare species, and should only be used when the species number of the community is known and the collected specimens are a truly random sample of the original community (Pielou, 1966).

A site containing a thousand species might not seem particularly diverse if 99.9% of individuals are of the same species (Purvis and Hector, 2000). It is therefore necessary to measure the extent to which individuals are distributed evenly among species, defined as **evenness**. However, evenness measures are less commonly used than heterogeneity and their use is generally not recommended (Etter, 1999). Unless the number of species in the community is known, they always overestimate evenness. The calculation of an evenness measure consists of the scaling of a heterogeneity index to its maximum possible value (when each species in the sample is equally abundant):

$$\text{evenness} = D/D_{max} \qquad (3)$$

where D is the observed index of species diversity and D_{max} is the maximum possible index of diversity given S species and N individuals. Evenness measures range from 0 to 1. In a study of assemblages of Jurassic bivalves Aberhan and Fürsich (2000) calculated the evenness of guilds using equation $E = 1/\Sigma p_i^2$, p_i is the relative abundance of the ith guild.

As stressed by Brown (1988), the nature of the fossil record makes the application of any of these indices to extinct biotas hazardous. Thus, palaeobiogeographers simply quantify diversity in terms of S, the total number of taxa censused at selected localities. Due to frequent taxonomic problems for geologically old groups and those lacking modern representatives, authors generally use generic instead of specific diversity. This widespread practice has been questioned by Budd *et al.* (1994), who showed that during the Plio-Pleistocene the generic richness of Caribbean reef coral fauna fell although species richness remained relatively constant. These authors concluded (1994, p. 971) that 'different patterns prevail at the species and genus levels, and genus-level patterns cannot be used to approximate species-level patterns'. In contrast, however, O'Dogherty *et al.* (2000) have shown that there are no substantial differences in patterns of generic and specific biodiversity in Jurassic ammonites, except in cases of suspected oversplitting, a point already stressed by Hallam (1987). Benthic and nectonic organisms seem to display different patterns of generic and specific diversity, and care must be taken when using conclusions based on patterns of generic diversity as a proxy for specific diversity.

Biologists and palaeoecologists commonly use three types of diversity (α, β, γ), defined by Whittaker (1972).

- *Alpha diversity* – Measures the level of complexity within the community. It reflects the scale at which species are dividing ecological resources. It is high if species are numerous and their relative abundances are similar, i.e. high equitability means high values of alpha diversity and high complexity. In palaeoecology, this type of diversity is often equated with the concept of richness, or with *S*, the total number of species, or with the number of genera (Sepkoski, 1988).
- *Beta diversity* – Between two communities, 1 and 2, beta diversity is expressed by:

$$D\beta = D\alpha_{12} - 0.5(D\alpha_1 + D\alpha_2) \qquad (4)$$

  $D\alpha$ being calculated by equation (2). In contrast to alpha diversity, beta diversity expresses the difference between communities. It can be considered as a measure of the amount of turnover in species composition along environmental gradients, reflecting the extent of habitat selection and specialization (Sepkoski, 1988). According to Blondel (1979), beta diversity is an index of similarity because it varies between the extreme values 0, when the two communities are identical, and 1, when they have no species in common.
- *Gamma diversity* – Measures the diversity of 2, 3 . . . *n* mixed biotopes. It increases as the diversities measured for the communities of a large-scale area (e.g. a continent) are progressively cumulated. It can be used to measure differences in diversity between continents or to define patterns of diversity change in an area through time. Sepkoski (1988) recalls that Whittaker (1977) reserved gamma diversity, which he called 'landscape diversity', for the combination of alpha and beta values and delta diversity for between-region, or provincial, differentiation.

Brown (1988) stressed the difficulties inherent in dividing an apparently continuous range of spatial variation into discrete units. For him, diversity increases with the size of the area sampled. All measures of species richness are highly dependent on the spatial scale of sampling.

Rosenzweig (1995) dismissed these distinctions because gamma diversity 'is no more than the number of species in a region' and 'the slope of the mainland species–area curve is one measure of beta diversity' (1995, p. 33).

However, the differentiation of these three types of diversity has proved to be useful in macroevolutionary studies. For example, Sepkoski (1988) noted that the increase of alpha diversity from 50% to 70% in offshore marine environments during the Ordovician (Bambach, 1977) does not account for the 300% rise observed in the global generic diversity. The 'missing diversity' could be attributed to measured changes in beta and gamma diversities. The remaining global diversity has to be sought in hidden sources of beta diversity because the lack of increasing provinciality or endemism from the Cambrian to the Ordovician points to a subordinate role for gamma diversity (a measure of biogeographical differentiation).

7.2 Communities and guilds

The term 'community' identifies a complex biological system recurrent in time and space that has collective properties of higher rank than the population, e.g. trophic structure, diversity, stability. The community is therefore not a fortuitous assemblage of species but corresponds to a level of organization with properties different from those of populations. The cohesion of the community is a result of the complementary needs and functions of its different components, i.e. the species.

Palaeontologists are concerned with the recurrence of communities in time and space. This can be particularly important where palaeobiogeography is used for palaeogeographical reconstruction. For example, reefal communities occur throughout the Phanerozoic and in different parts of the globe indicate intertropical latitudes. Taxonomic coherence is not required in the palaeoecological analysis because this generally investigates the functional properties of the elements of this kind of community. On the contrary, taxonomic relationships are important in palaeobiogeography because they may point out the geographical dimension of evolutionary pathways, as also reflected in cladistic biogeography.

Within a community, a **guild** corresponds to a small group of species that share, with different modes, the same type of resource. The components of a guild are not taxonomically related but show ecomorphological ressemblances. This concept is commonly used in ecology and, at a coarser scale, also in palaeontology (Bambach, 1985, 1993; Aberhan, 1994), although it is impossible to define guilds with the same accuracy as in ecology. Nevertheless, within trophic categories such as suspension-feeders, detritus-feeders, carnivores and herbivores, coarser groups of ecomorphologically similar organisms can be differentiated. The concept of guild is a useful means of studying the process of niche segregation during adaptive radiation. This process has obvious biogeographical implications that are reflected in the spatial patterns of diversity. Relationships between trophic resources and guild-types may aid interpretation of the biogeographical history of particular regions. An interesting example is that of the oceanic assemblages of echinoderms, whose distribution and diversity vary with both depth and primary production. Roux (1982) has shown that in the NE Pacific the diversity of the bathyal holothurian epifauna is controlled by nutrient supply. Species numbers increase with depth, from the continental slope to the ocean floor. Holothurians are represented in this area by detritus-feeding species, but beyond the Juan de Fuca ridge diversity decreases and suspension-feeders dominate. Roux also suggested that comparable characteristics of photic and continental primary production explain the similarities of modern NE Pacific and NE Atlantic asteroids and holothurians. The colonization of the ocean floors by stalked crinoids (suspension-feeders) is the result of a long and complex history involving adaptive strategies that operated along oceanic ridges and were favoured by hydrothermal trophic resources.

The study of both communities and guilds in palaeoecology is faced with practical problems reflecting the process of fossilization. First, soft-bodied organisms typically leave no trace of their occurrence in a community. Second, even if the effects of taphonomic processes can be detected, the preserved components of an assemblage may correspond to elements of different communities transported to the site through different biostratinomic (before burial) pathways. Taphonomic condensation (*sensu* Fernandez-Lopez, 1991) may concentrate organisms belonging to different ecological assemblages in a single bed. Taphonomic analysis is therefore indispensable to avoid the fruitless study of spurious assemblages.

7.2.1 Faunal spectra

A quick and practical means of studying some aspects of biodiversity, particularly the composition of a fossil assemblage and the relative abundance of its constituent taxa, is the construction of faunal spectra. This method was employed in the 1960s by Ziegler (1967) and has gained wider use subsequently. The faunal spectrum has been defined (Cecca, 1992) as a cumulative picture in space (reflecting the sinking of

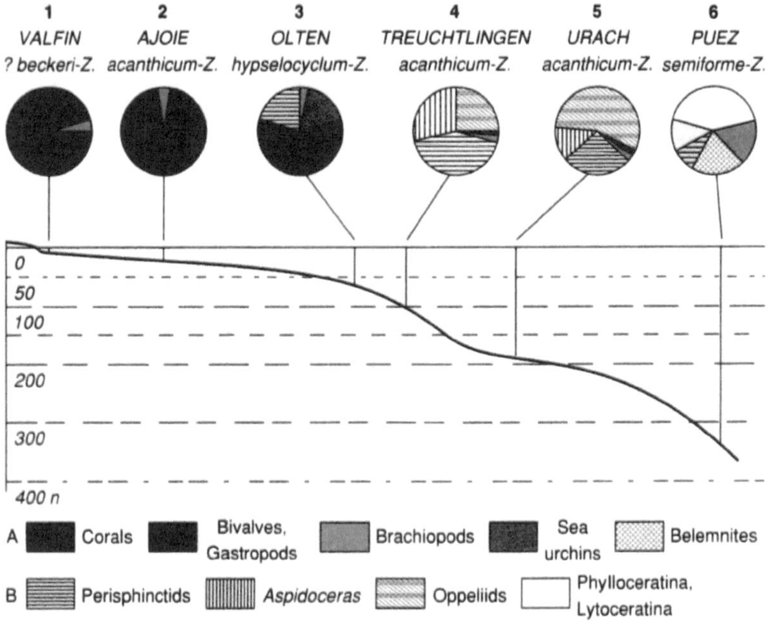

Figure 7.1 Faunal spectra of communities of ammonite zones at various localities. The relative frequencies of benthic groups suggest a bathymetric gradient. The ages are Late Kimmeridgian for spectra 1, 2, 4 and 5, Early Kimmeridgian for spectrum 3 and Early Tithonian for 6. In A the segments correspond to taxa other than ammonites, whereas in B they indicate different ammonite groups.

Source: Ziegler (1974). Reproduced with permission of BRGM.

shells of organisms that lived at different levels of the water column plus drifted shells) and time (depending on the rate of sedimentation).

Faunal spectra provide a measure of ecological changes in the structure of faunal assemblages along environmental gradients, stretching for example from shallow to deep shelf and basin (Fig. 7.1). However, the value of these data depends heavily on the preservation of fossil communities. Preliminary taphonomic analysis of the fauna is thus indispensable. Faunal spectra can be represented in different ways; the most widely used are pie charts and bar charts, subdivided in both cases into sectors assigned to the individual taxa observed in the community studied. The angle of each sector of the pie chart, or the length of each bar, varies in proportion to the relative abundance of the taxon within the community. Faunal spectra may be used to estimate palaeodepth and to assess the relative importance of a guild within a community, provided that the trophic characteristics are known.

7.3 Biogeographical patterns of biodiversity

The literature on biodiversity patterns gives more emphasis to ecological, mainly biotic, factors than to historical factors. However, the existence of geographical biodiversity gradients (both latitudinal and longitudinal) and their importance for macroevolutionary patterns (e.g. the biogeography of mass extinctions and following recoveries) strongly suggest that historical and environmental factors have been just as important as, if not more important than, biotic and ecological ones.

7.3.1 The hollow curve

The biogeographical frequency of distribution is a quantitative measure of the distribution of taxa in increasing numbers of regions. The plot in Fig. 7.2 shows that the proportion of rudist species that are pandemic is very small, whereas endemic species are common. According to Flessa and Thomas (1985), who have studied the biogeographic distribution of Modern bivalve genera, the shape of such frequency distribution plots reflects the interaction between the probabilities of speciation, extinction and range expansion. Taxonomic bias does not seem to generate a concave distribution. Despite the few exceptions, the distinctively concave biogeographic frequency distribution, the 'hollow curve', seems to be characteristic of a variety of animals and plants. Similar curves have been generated in studies of the distribution of species among genera and of both species and genera among areas. Willis (1922) explained the prevalence of the hollow curve among diverse groups by his hypothesis of 'age and area'. Endemic species are geologically young whereas widespread species are relatively old. Note that the plot in Fig. 7.2 does not include time.

Models based on random rates of speciation and extinction do generate the hollow curve, and thus support Willis' hypothesis. If biogeographical regulation of evolutionary rates exists, the shape of the biogeographical frequency distribution also depends on range expansion. Thus, the simulation carried out by Flessa and Thomas (1985) showed that the frequency distribution is not sensitive to the probabilities of extinction and speciation. On the contrary, the hollow curve of geographical distribution simply results from the increasing probability of range expansion with increasing geographical range. This conclusion is biologically justified by the common association of wide

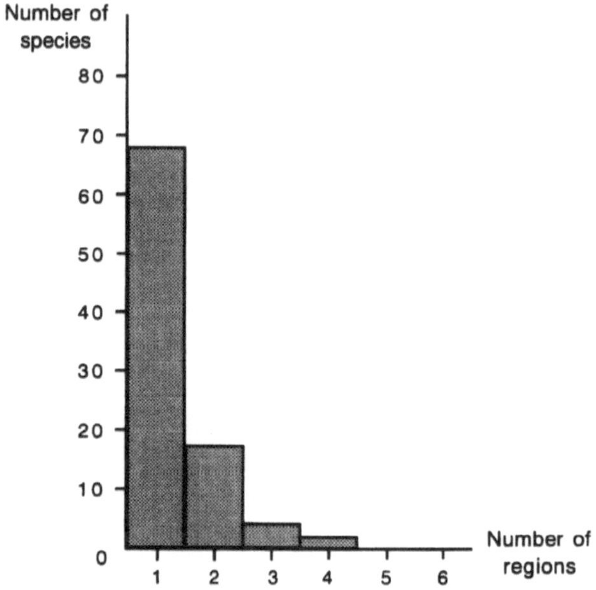

Figure 7.2 Biogeographic frequency of distribution of Campanian–Maastrichtian rudist species in six palaeobiogeographic regions (SW France, Tunisia, SE Spain, Apennines and SE Italy, Sicily and Tunisia). The plot shows that most species occur in only one region and that the numbers of species present in increasing numbers of regions decrease rapidly. No species are distributed in more than four areas. Author's original plot based on data published by Philip (1985).

geographical range with high dispersion ability (see Chapter 6), demonstrated by larval biogeography. The fossil records of molluscs (Bretsky, 1973; Hansen, 1980; Jablonski, 1980; Hoffmann and Szubzda-Studencka, 1982) and brachiopods (Boucot, 1975a) indicate that widespread species tend to have longer geological ranges. The probability of extinction appears to decline with increasing geographical range, as demonstrated by the evolutionary implications of the different types of larval development.

7.3.2 Latitudinal gradients of biodiversity and evolutionary implications

One of the most obvious aspects of the geographical distributions of both marine and terrestrial organisms is the existence of latitudinal diversity gradients. Biodiversity is higher in the tropics and decreases towards the poles (Brown, 1988; Rosenzweig, 1995; Gaston, 2000). This pattern is recognizable in the record of the majority of fossil invertebrate groups, at least since the Mesozoic, when plate tectonic reconstructions are most reliable. This supposedly perennial gradient generates strong debates about the relationships between species diversity, evolutionary rates and the biogeographic dimension. For example, Rosenzweig (1995) considers this gradient to be a time-invariant feature of large-scale biogeography due to the greater area of the Tropics. However, a number of mechanisms that do not involve biogeographical factors have been suggested as controls, as area alone seems insufficient to explain latitudinal biodiversity gradients (Gaston, 2000). In particular, climate and climate changes play a fundamental role in controlling the distribution of the diversity (O'Brien, 1993, 1998).

Before discussing the possible causes of the latitudinal diversity gradient we will examine patterns of richness in species and genera of recent and past marine invertebrates.

Roy *et al.* (1996) analysed a large data set for eastern Pacific marine molluscs and were able to show the distribution of species within higher taxa. In particular, the number of families is higher at low latitudes and the number of genera that have only a single species shows a strong latitudinal gradient. At all latitudes, the frequency distributions of species within genera plot as a hollow curve. This differs from the hollow curve described above (plotting genera against biogeographic regions) and shows that a high proportion of genera are species-poor (Fig. 7.3). The pattern is emphasized at low latitudes because the absolute number of species-poor genera is considerably higher, biodiversity being generally higher at low latitudes. These genera might always have been species-poor during their geological life or be relicts of formerly species-rich taxa. Variations in species richness among higher taxa can be attributed to abiotic factors such as the availability of nutrients, but the large number of monotypic genera and species-poor families might also reflect an evolutionary dynamic, with young and species-poor taxa concentrated in the Tropics (intended as a 'cradle').

Roy *et al.* (1998, 2000) have shown that at the species level there is a marked latitudinal gradient of decreasing diversity from low to high latitudes of living marine prosobranch gastropods in the Western Atlantic and Eastern Pacific Oceans. Epifaunal and infaunal bivalves (with the exception of the protobranchs) show similar latitudinal trends in the latter area. This result, which has been confirmed by Crame (2000), is in contrast with the hypothesis that a latitudinal diversity gradient occurs in the marine epifauna but not in the infauna (Thorson,

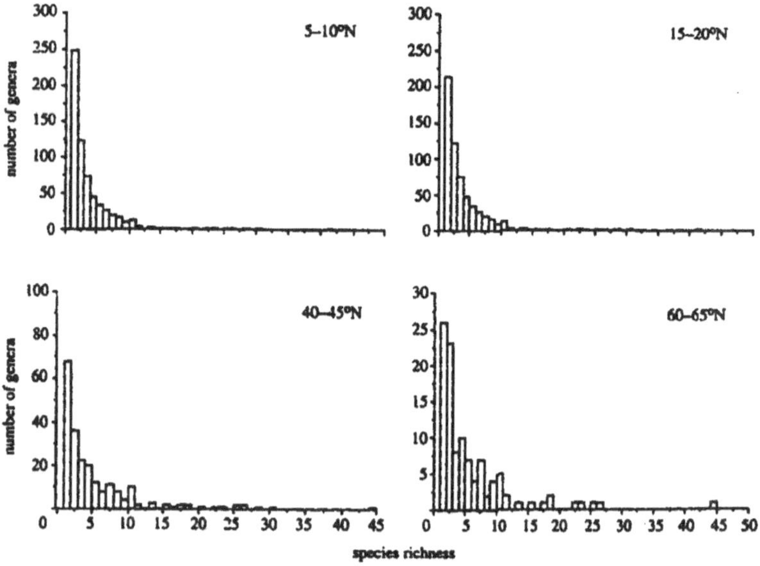

Figure 7.3 Frequency distribution of species (horizontal axis) within genera of eastern Pacific molluscs at four latitudinal, tropical and temperate, bands. Note the 'hollow curve' pattern.

Source: Roy *et al.* (1996). Reproduced with permission of the Royal Society.

1957). The species numbers for both bivalves and prosobranch gastropods rapidly increase from the Equator and peak at around 10°N, where they are relatively constant to 23°N (the northern edge of the Tropics) for the bivalves, and around 30°N for the gastropods. Diversity then falls by about 50% in both groups. Among the bivalves, the ratio of infaunal to epifaunal species shows an overall positive trend with latitude (Fig. 7.4). Interestingly, the gentle decline of the values of this ratio in the Tropics and their slight increase between 35° and 50° (Roy *et al.* consider the peak at around 65° unlikely to be biologically meaningful) suggest an opposite trend in past geological periods such as the Late Jurassic (Crame, 1996) and Late Cretaceous (Roy *et al.* quote D. Jablonski's unpublished data).

Further evidence of latitudinal gradients can be derived from age–frequency distributions of taxa, particularly at the genus level, in the tropical and nontropical areas. Boucot (1975a) plotted the numbers of genera/subgenera of hippuritid rudists and hermatypic and ahermatypic corals against duration classes and stressed that taxa occurring

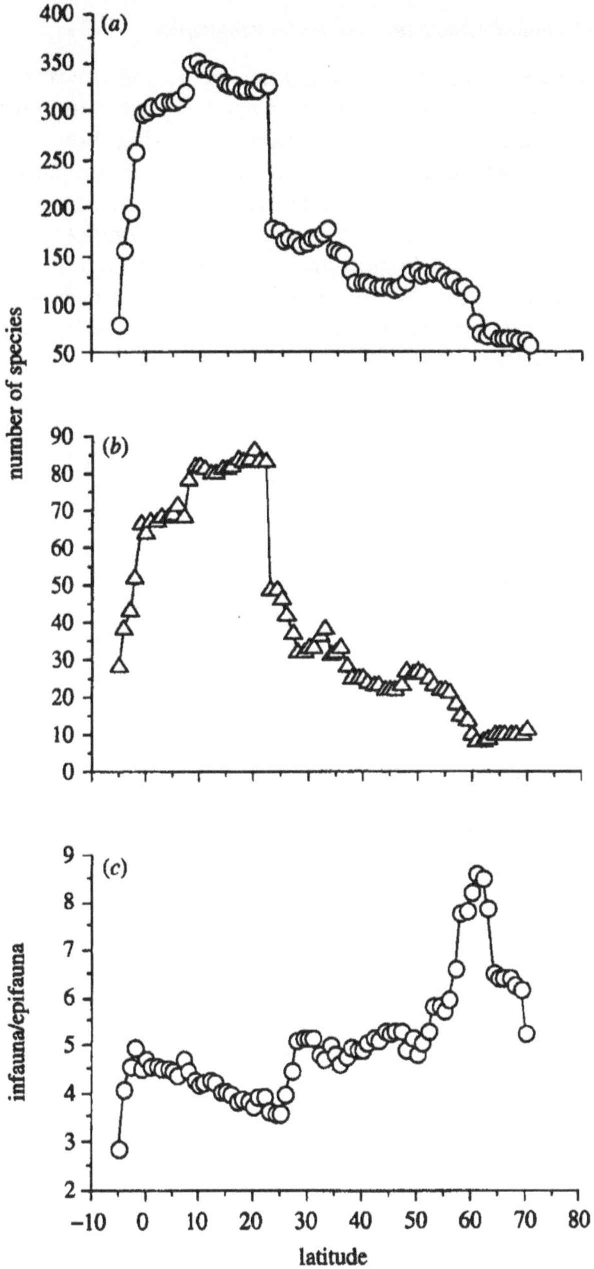

Figure 7.4 Latitudinal diversity gradient of (a) infaunal and (b) epifaunal north-eastern Pacific bivalves; (c) the latitudinal trend in the ratio of infaunal versus epifaunal bivalves from the same area.

Source: Roy *et al.* (2000). Reproduced with permission of the Royal Society.

over a wide area have longer stratigraphic ranges and vice versa. He also indicated that taxa with a low level of phenotypic variation tend to be both geographically restricted and rapidly evolving. Boucot's conclusions are valuable, but cannot represent a rule. They are partly contradicted by some ammonoid taxa. For example, the ammonoid suborder Phylloceratina is characterized by slowly evolving (long-ranging) species with low phenotypic variation and wide (but mainly tropical) distributions. Rosen (1984) also cautioned that some old reef coral genera have narrow geographical ranges but have relatively large numbers of constituent species, comparable with more recent species-rich genera.

Flessa and Jablonski (1996), based on an analysis of geological age and geographical distributions of bivalve genera, and a review of data on corals, barnacles, and planktonic foraminifers, concluded that the evolutionary productivity of the Tropics extends across levels of taxonomic hierarchy. Geological age–frequency distributions are obtained by plotting the percentage of taxa that fall within a series of duration classes (Fig. 7.5). The distribution for cosmopolitan bivalve genera, which occur in the Atlantic, Pacific and Indian oceans, shows its highest modal value in the 60–70 m.y. age class, a secondary mode in the 20–30 m.y. age class and a considerable number of taxa dating back more than 100 m.y. In contrast, the curve for less widespread genera, which do not occur in all three oceans, shows a peak in the 10–20 m.y. age class. This result is consistent with the well-documented greater geological duration of widespread taxa which have greater ecological tolerances, greater dispersal ability and greater global population sizes. If long geological durations are the result of widespread distributions, higher origination rates in the Tropics are suggested by more frequent endemic distributions of tropical taxa characterized by shorter stratigraphical ranges.

The relationship between longevity and latitudinal gradients of diversity has been confirmed for Neogene planktonic foraminifers by Parker *et al.* (1999) (Fig. 7.6). As a rule, cosmopolitan species are again characterized by longer durations, but these are essentially polar and subpolar species. Short-lived species are more numerous in tropical–temperate latitudes. The conclusion is that species of polar and subpolar areas have lower speciation rates than those of lower latitudes.

However, deviations from the expected pattern of declining diversity from low to high latitudes, based on the observed modern patterns, are documented. For example, Doyle (1987, 1992) has recorded an increase in the generic diversity of belemnites in the Atlantic from the Boreal province (North-West Europe and East Greenland) to the Arctic province (Svalbard, Siberia, Arctic Canada, etc.) during the

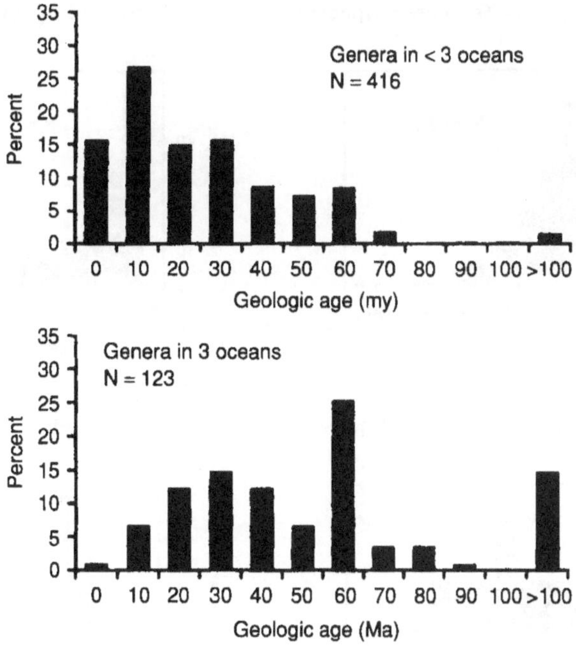

Figure 7.5 Age–frequency distribution of (upper) bivalve genera not found in all three oceans (Atlantic, Indian and Pacific Oceans) and more widespread genera that occur in all three oceans (lower).

Source: Flessa and Jablonski (1996). Reproduced with permission of University of Chicago Press.

Middle Jurassic–Early Cretaceous interval. In the Arctic province belemnite diversity was comparable to that of Tethys in the same interval.

Gaston (2000) stressed that the latitudinal gradient in species richness is a gross abstraction because any underlying pattern is concealed by variation in species richness reflecting other positional (longitude, depth, etc.) or environmental (currents, salinity, etc.) variables. Latitude *per se* is not a determinant of species richness, but only a correlative of a variety of causal environmental factors. However, the contrast between the high diversity of many tropical floras and faunas and the low diversity of their high-latitude counterparts is certainly striking. This explains why so many biologists and palaeontologists feel that the latitudinal gradient in species richness is an intriguing pattern that may be the expression of macroevolutionary mechanisms.

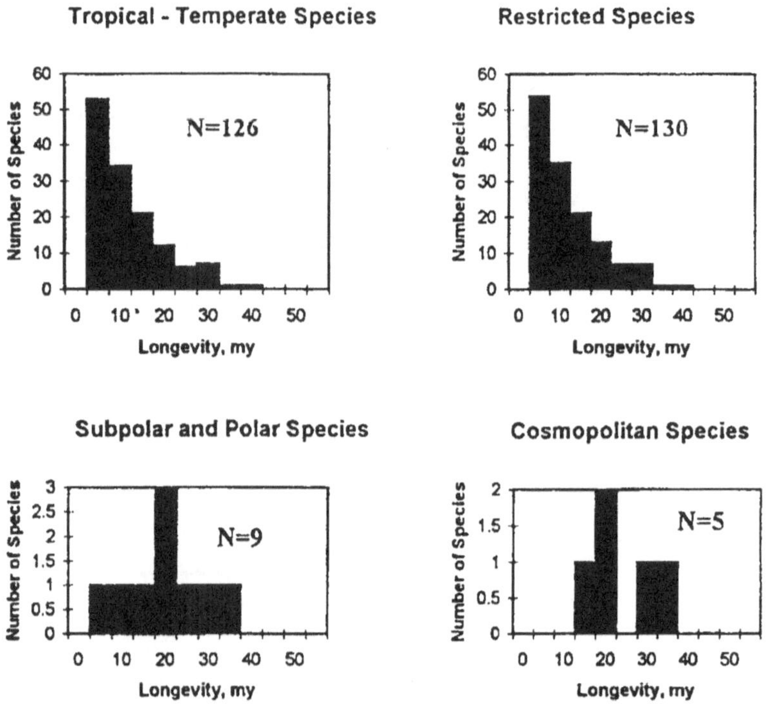

Figure 7.6 Distributions of species longevity in Neogene planktonic foraminifers, showing latitudinal differences (left histograms) and biogeographical patterns expressed by number of latitudinal zones over which a species ranges (right histograms).

Source: Parker *et al.* (1999). Modified by permission of Elsevier Science.

7.3.2.1 Are the Tropics a 'cradle' or a 'museum'?

The Tropics are extraordinarily rich in species, higher taxa and evolutionary novelties, and appear to gather, on average, younger taxa. The evolutionary implications of the latitudinal diversity gradient shown by many taxonomic groups have been much debated, giving rise to the famous fundamental question of whether the Tropics are 'a cradle or a museum' (Stebbins, 1974). Low-latitude richness may be explained by high origination rates, with the Tropics operating as a diversity pump, by generally low extinction rates, with the Tropics operating as a diversity accumulator, or by some combination of the two (Jablonski, 1993, with references therein).

Arguments in favour of high origination rates in the Tropics are supported by the first appearances of post-Palaeozoic marine orders, whether defined latitudinally (with estimations of palaeolatitudes deduced by palaeomagnetism) or biogeographically (on the basis of tropical geoecological assemblages such as abundant dasycladalean algae, large benthic foraminifera). Jablonski (1993) was able to demonstrate higher origination rates in tropical than in nontropical seas. However, apparent patterns of first origination might be determined by preservational factors or by more frequent sampling intensities. Potential sampling artefacts might also arise from the decreasing area of each belt with increasing latitude. Taking into account members of poorly preserved orders (lightly mineralized, fragmented or disarticulated remains) and measures of sampling intensity across latitude, and then using adjustment procedures to evaluate both sampling and preservational factors, Jablonski reached two important conclusions (Fig. 7.7):

1 despite the larger total area of the tropical regions, well-preserved orders are more poorly sampled
2 the proportion of well-preserved orders originating in tropical seas is greater than expected.

These results suggest that, at least since the Mesozoic, novel body plans originate most frequently in tropical taxa but do not mean that all taxa have tropical origins. Many marine taxa of a variety of taxonomic levels clearly originated in high latitudes.

Widespread taxa have longer stratigraphical distributions and tropical taxa tend to have more restricted biogeographical and stratigraphical distributions. Therefore high-latitude taxa should be more cosmopolitan and long-ranging, thus raising the question of the evolutionary role of high-latitude areas. In particular, it would be interesting to compare the evolutionary impact of combined lower origination rates and lower extinction rates recognized in polar and subpolar regions with rates of the tropical regions. A tropical molluscan clade contains on average three times as many species as a cold temperate/polar one, but Crame (1996) asked whether tropical clades in fact have higher speciation and extinction rates than those of high latitudes and polar areas. He showed that there is a statistically significant relationship between the number of species in a clade and its age, and that this does not differ between tropical and cold temperate/polar clades. The most diverse clades, besides all being tropical, are also the oldest. Crame also emphasized the fact that not all tropical clades have radiated

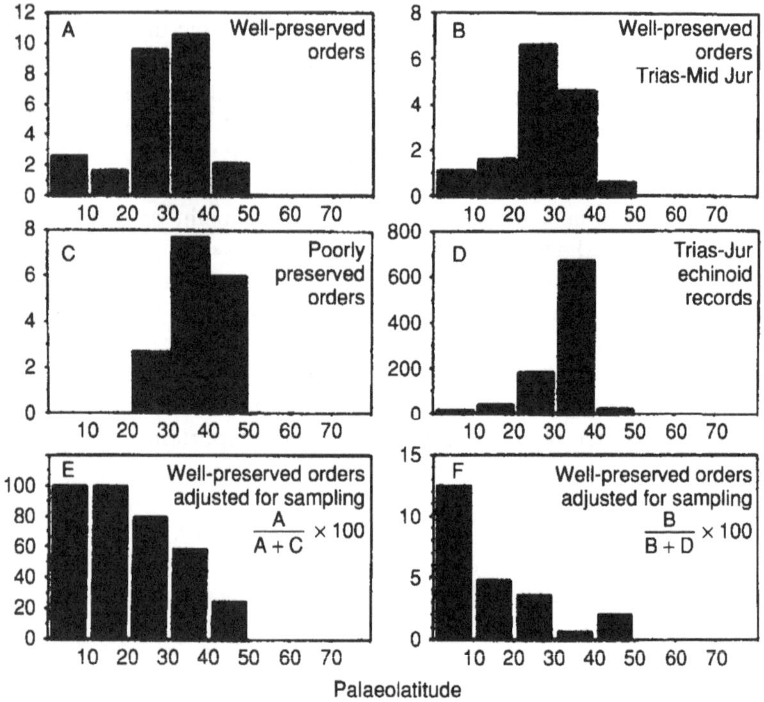

Figure 7.7 First occurrences of orders of post-Palaeozoic marine benthic invertebrates, grouped in 10° palaeolatitude belts (N and S plotted on same axis). A, well preserved orders, Triassic–Recent; B, Triassic–Bajocian (the interval that includes more than half of the ordinal originations); C, poorly-preserved orders (Triassic–Recent); D, Triassic–Bajocian echinoid records, calculated as the number of localities from which each species is recorded; E, well-preserved orders from A, adjusted for preservation and sampling using the distribution of poorly-preserved orders in C; F, Triassic–Bajocian well-preserved orders in B, adjusted for sampling using the distribution of echinoid records in D.

Source: Jablonski (1993). Reproduced by permission of the author and *Nature*.

rapidly and not all cold temperate clades have radiated slowly. This implies that tropical clades do not radiate at a significantly faster rate. The fact that many tropical habitats are older than many temperate ones may explain the higher number of species in tropical clades. MacLeod and Keller (1994) suggested that high-latitude regions served as refuges at the K–T boundary and as a post-extinction source for oceanic plankton. Crame (1993), however, suggested that some groups may have accumulated in the Tropics. In this case, the Tropics would not produce as many evolutionary novelties as is usually thought.

The idea of the 'cradle' recalls the general idea of a 'centre of origin' discussed in Chapter 3. The higher diversity and younger ages of Indo-West Pacific (IWP) reef corals have been explained by Rosen (1984, 1988c) as a relatively recent phenomenon reflecting vicariance processes induced by Neogene glacio-eustatic fluctuations. The age-and-area patterns which Flessa and Jablonski (1996) use to support the idea of the Tropics as a 'cradle' imply that diversity in the IWP Focus is older than Neogene. Flessa and Jablonski (1996, p. 388) state that 'neither the diversity gradient nor the age differences appear to be a strictly Neogene phenomenon'. However, Wilson and Rosen (1998) have recently demonstrated the reality of a 'Paleogene gap' in IWP areas, and this supports a Neogene origin of the biota. The role of climate has been discussed at length by Crame (1993) to explain, at least in part, bipolarity. Crame observed that just as range expansions across the Tropics are axiomatic to the formation of bipolar distributions, so range retractions into these areas, enhanced by climate warming, may be crucial to the establishment of high-diversity foci in low latitudes by a slow but steady diffusion of taxa.

The evolutionary model for reef coral diversity, and the problem of bipolar distributions since the Mesozoic, challenge the model of the Tropics as a 'cradle' and extend the scope of palaeobiogeographical and palaeoecological research. Once more, evidence from other groups is needed.

7.3.2.2 Causal hypotheses

Several different hypotheses, which in most cases are not mutually exclusive, have been put forward to explain latitudinal diversity gradients.

Communities tend to become increasingly diverse with time. Thus, Jablonski's results (discussed above) suggest that high diversity in the Tropics is due to the production of evolutionary novelties since the Triassic. Spatial heterogeneity is higher in the Tropics. Competition and predation are strong in highly specialized tropical communities and may favour the development of diversity. Rapoport's rule (Rapoport, 1975; Stevens, 1989) suggests that the latitudinal diversity gradient follows as a result of both the contraction of species' distributional range and progressively widening regional species pools (richer biotas) toward low latitudes. In other words, compared to high latitudes the richer tropical biotas result from both an increased subdivision of ecological niches and the larger surface of the Tropics.

The species–area relationship is considered to govern the observed present-day latitudinal diversity because continental areas are concentrated in low latitudes and tropical oceans cover a larger surface area than those in other climatic zones (Rosenzweig, 1995).

A warm, wet and stable climate favours the stability of trophic resources, allowing a high number of species to coexist. The environmental stability of the tropics is widely considered the most important factor in explaining the high diversity levels.

Species–energy relationships are thought to be important in determining latitudinal gradients of biodiversity. Greater availability of energy is assumed to enable a greater biomass to be supported in an area. This in turn enables more organisms to coexist, and thus more species at abundances that enable them to maintain viable populations. Species richness therefore increases with available energy. Ambient available environmental energy is usually estimated from models or indirectly from other variables, and is often used interchangeably with 'net primary productivity' (Gaston, 2000). Solar energy input and its relationship with productivity and trophodynamic processes must be taken into account. Species–energy hypotheses do not take into account solar energy alone because energy is optimized by climatic conditions, which therefore play an important role in controlling diversity (O'Brien, 1993, 1998).

Finally, historical events have been invoked too, as mass extinctions have repeatedly upset the system (Brown, 1988).

Roy *et al.* (1998) used the biogeographical patterns of living marine prosobranch gastropods of the Western Atlantic and Eastern Pacific Oceans to test these hypotheses. Higher median latitudinal ranges in low rather than high latitudes negate Rapoport's rule. There is no support for the species–area effect, with a lack of significant correlation between shelf areas and diversity. However, the latitudinal diversity gradient depends on environmental variables related to solar radiation, such as temperature, productivity and seasonality, which are likely to influence evolutionary dynamics (Roy *et al.*, 1998, 2000). This conclusion is evident in the framework for the economic theory of biological evolution (Vermeij, 1995), although Roy *et al.* acknowledge that the mechanisms linking these variables to evolutionary rates are poorly understood. According to Flessa and Jablonski (1996), who consider the Tropics a cradle and not simply a museum for biodiversity, origination and extinction rates are higher in tropical than in non-tropical regions and therefore produce a biota rich in young species.

High origination rates result from factors such as the area, complexity and resource stability of the Tropics, whereas high extinction rates may be a consequence of narrow ecological tolerances, restricted geographic distributions, or delicate nutrient regimes.

Observation of living faunas and floras is the most used method of testing these hypotheses. It provides both easy access to the large neobiogeographical database and direct measurement of relationships between biotas and environmental factors. The major risk is the potential for underestimation of the effects of historical events in shaping present-day patterns. For example, if the species–area relationship is really as important as we believe, we should take into account the fact that the distribution of lands and oceans across latitudinal belts has changed during the Phanerozoic (Wyatt, 1993).

The distribution patterns of recent zooxanthellate corals relative to both latitude and mean annual sea surface temperatures clearly show that at the same latitude and in the same temperature range the highest Atlantic diversities are always lower than the highest IWP values. The difference in diversity is approximately 1 to 4 at the generic level (Rosen, 1984). This differentiation can be taken to argue that the diversity pattern is not linked to temperature or, at least, not to temperature alone. The fauna at any one locality results from two distinct influences, a latitudinal influence (implying climatic factors) and a longitudinal effect (implying independence of major climatic belts). For example, in the eastern parts of the present ocean basins cold boundary currents cause coral reefs to be restricted to a much narrower latitudinal zone than on western ocean margins. However, long-term events cannot be disregarded. A sharp decline in the total number of reef coral genera in the Caribbean is correlated with the Late Pliocene cooling and the effects of Pleistocene glaciations. In addition, the closure of the Panama Isthmus affected the shaping of the Pacific longitudinal gradient by leaving some genera of Caribbean origin isolated in the eastern Pacific (Budd *et al.*, 1994).

7.3.2.3 What about polar regions?

The latitudinal diversity gradients shown by living molluscs and reef corals suggest a relationship between diversity and energy-related variables. Although the fossil record generally provides evidence of the enduring character of such a gradient, research on Mesozoic high-latitude assemblages suggests that long-term changes in climate and geography should also be considered. The model of the Tropics as a cradle may implicitly suggest that taxa originate there and spread to the poles, because tropical marine invertebrate faunas seem to be characterized by geologically younger taxa. However, as stressed in the discussion of the ideas of a 'centre of origin', neither the predominance of geologically young taxa nor the greatest diversity of living species in an area is proof of an origin there. They may be the result of adaptation. The generally equable climate during the Mesozoic, with less pronounced gradients than at present, might be seen as an indication that distributional changes with macroevolutionary effects occurred at the end of the greenhouse phase. Origination in high latitudes has been suggested for some mollusc groups by Skelton *et al.* (1990) and Crame (1992, 1993), and a bipolar distribution for some Early Jurassic bivalves has been demonstrated by Damborenea (1993). From a vicariant perspective, polar taxa may represent relicts of formerly cosmopolitan clades (Crame, 1992). The taxonomic and stratigraphic analysis of bivalves carried out by Crame (2000) shows that the steepest latitudinal gradients are associated with the youngest clades, lending support to the idea that the Tropics have served as a major centre of evolutionary innovation. Crame also suggests that these patterns are most likely to be the product of primary radiations because it may take some taxa tens or even hundreds of millions of years to diffuse from the Tropics to the poles, which have not yet been reached by the youngest bivalve clades. Finally, the steepness of the latitudinal gradients may be a relatively young feature of the biosphere because the tropical ocean has existed at least since the beginning of the Mesozoic, whereas extensive cold polar seas date back to no more than the Eocene–Oligocene transition (Keller, 1983; Crame, 1993).

7.3.3 Longitudinal gradients and causal hypotheses

These gradients have been poorly studied, despite their potential long-term significance and their importance in palaeogeographical reconstructions. As Flessa and Jablonski (1996, p. 389) pointed out,we should recognize that 'Longitudinal comparisons among oceans have been performed less often than latitudinal ones, and the patterns arc even less understood'. Nevertheless, longitudinal gradients are recognizable in the fossil record. Like reef corals, larger foraminifers show a west-to-east decline in diversity in the Pacific (Belasky, 1996). This longitudinal gradient has been recorded in geological periods since the formation of Pangaea during the Late Carboniferous–Early Permian, giving impetus to estimates of palaeolongitude on the basis of the eastward diversity decrease in the IWP Ocean (see Chapter 8). The age, area patterns (Stehli and Wells, 1971), and eastward decrease in numbers of reef coral genera (Rosen, 1984) and larger foraminifers (Belasky, 1996) may suggest radial diffusion outwards from the focus (Atlantic or IWP), and have given considerable impetus to the idea of the IWP as a 'centre of origin'.

As already indicated, cold currents may lower both distribution and marine biodiversity off western continental margins. A remarkably similar pattern is shown on maps of Tethyan palaeoenvironments plotting occurrences of reef corals (Dercourt *et al.*, 1993). Over a series of time-intervals reefal formations were mostly developed off the eastern sides of the continental masses. In addition to the persistent Pacific gradient, longitudinal diversity gradients that formed following the progressive westward opening of Tethys have been recorded in a variety of groups. These gradients developed after the break-up of Pangaea and before the closure of Tethys, the most environmentally significant geological phenomena to have shaped the Earth during the Mesozoic and Cenozoic. A few examples serve as illustration.

The highest diversity in Jurassic ammonites is recorded in Mediterranean areas of Tethys, with both eastward and westward decline from the focal region. However, the eastward decline is more speculative for the Jurassic because of the latitudinally oblique distribution of eastern Tethyan continental margins, and the removal of potentially informative outcrops as a result of the Alpino-Himalayan orogenesis (Cecca, 1999). Tectonic movements are also important in the eastern Mediterranean during the Lower and Middle Jurassic (Dommergues *et al.*, 1987a, 1987b; Dercourt *et al.*, 1993). The decline in diversity of Cretaceous inoceramids from Europe and the Mediterranean Tethys eastwards can be inferred from the maps presented by Dhondt (1992). Changing levels of provincialism in rudist assemblages also followed a roughly longitudinal pattern stretching along Douvillé's 'Mésogée' (Philip, 1982, 1998; Masse, 1992). Complex longitudinal patterns have been shown for Pteropods. Their development in the Pacific during the Pleistocene coincided with and followed the emergence of the Panama Isthmus. Today the Pacific retains a poor diversity compared to both Atlantic and Indian oceans (Bernasconi and Robba, 1982).

The existence of longitudinal patterns may lead to the conclusion that the distance from a supposed centre of origin is the sole factor determining a decrease in diversity. Piccoli *et al.* (1986) and Piccoli and Sartori (1994), for example, have shown the relationships between the numbers of common species and interbasinal distances calculated on palaeogeographical maps along the shortest seaways. They consider that the longer the distance, the smaller is the number of common species, and quantify the faunal exchanges between basins as a function of interbasinal distances through a diffusion law. However, the longitudinal similarities and age-and-area patterns should not be translated directly into geographical distances because different factors may act along different distance vectors (Belasky, 1996).

Biogeographical constraints derived from longitudinal patterns can be applied to problems of palaeogeographical reconstruction and are discussed at length in Chapter 8.

7.4 Equilibrium theory and species–area relationships

Since the days of Darwin and Wallace, one of the most emotive themes in biogeography has been the study of the distribution of insular biotas. In our daily lives the word 'island' represents a portion of land encircled by water. In biogeography the meaning is extended and an island is thought of as a portion of space separated from surrounding portions of space through natural, ecological and/or geographical barriers. In this context a lake may be considered and treated as an island. Islands may therefore be divided into those within the sea and those within the land. Islands are discrete geographical entities that ecologists and biogeographers consider natural laboratories where a wide array of processes can be studied and quantified. One of the most important areas of interest is certainly the possibility of observing the dynamics of the island populations. This may be with respect to both natural barriers that limit their distributional areas and characteristics of the environments in which they live. Islands allow us to test models of speciation and their relationship with a geographical dimension. Terrestrial animal and plant populations of islands have peculiar characteristics, and special features have also been shown in populations of marine molluscs. Research on island biotas has important applications to programmes of conservation biology, and is the reason why studies of island biogeography are mostly devoted to terrestrial biotas. Discussion of these is not the object of the present book, and I refer the reader to the monograph on island biogeography by Whittaker (1998).

However, even for marine palaeobiogeography (on the assumption that marine faunal provinces can be treated theoretically and analysed as islands), the interest in islands derives from MacArthur and Wilson's (1963) **equilibrium theory of island biogeography** (ETIB). The central idea of this is the assumption that ecological systems represent an essentially balanced state of equilibrium. This has been questioned by some authors and challenged by Whittaker (1998), who proposed

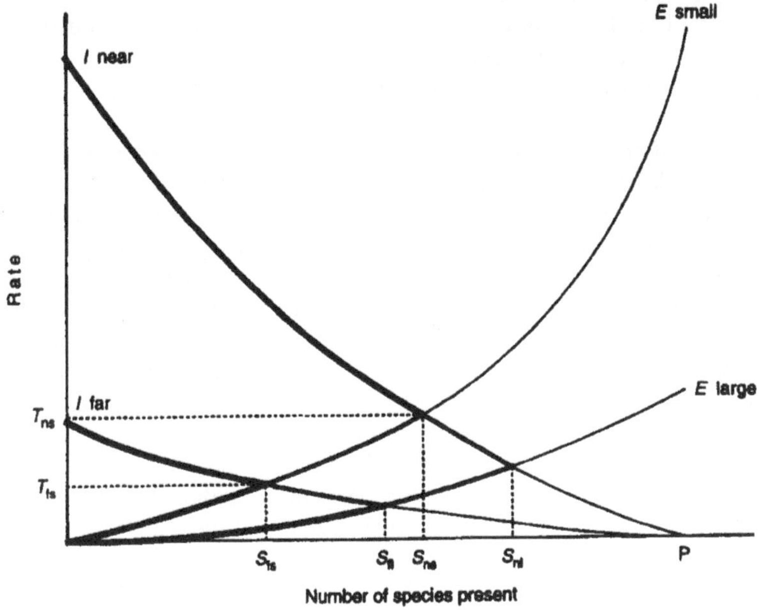

Figure 7.8 A version of MacArthur & Wilson's (1963, 1967) equilibrium model taken from Whittaker (1998). I = immigration rate; E = extinction rate; T = turnover rate; S = species number; P = mainland source pool; f = far; n = near; s = small; l = large. The figure shows how immigration rates are postulated to vary as a function of distance, and the extinction rate as a function of island area. The model predicts the species number present at equilibrium.

Source: Reproduced with permission of Oxford University Press.

a different view, that of the disturbed island ecology, based on the effects of events such as volcanic eruptions and hurricanes. Brown (1988, p. 73) also noted that in many systems the frequency and magnitude of disturbances is so important that species diversity and composition continuously change, even in local ecological communities that are open to relatively unrestrained immigration. Nevertheless, the ETIB has had a tremendous impact, providing quantitative bases to debates involving not only biogeography and ecological biogeography but also island ecology and conservation biology. MacArthur and Wilson considered three major elements in their theory (Fig. 7.8):

1 the species–area relationship
2 the island isolation, measured by the distance of an island from the mainland (intended as the source area)
3 the turnover rates as a function of area and distance.

Immigration rates vary as a function of the distance of the island from the mainland and extinction as a function of island area. It is worth noting that in the ecology of present-day communities the term 'extinction' has a meaning different from that used in palaeontology. It corresponds to 'local extinction', that is, the disappearance of a species from an island, and does not mean that the species has definitively disappeared from Earth. It is, however, important to keep in mind the concept of a metapopulation because whereas the 'extinction' of a local population of a species from an island or patch does not necessarily lead to the final extinction, the disappearance of a local population is taken into account in calculating the extinction rate.

There are numerous palaeobiogeographical studies of island fossil communities, particularly of the Pleistocene, but they are mostly concerned with terrestrial faunas.

In marine palaeobiogeography, the species–area relationship is particularly important to studies of the evolution of diversity during geological time relative to the number and size of faunal provinces.

7.4.1 The species–area curve: predictive biogeography

Biodiversity has increased from the Cambrian to the present time (Fig. 7.9), but this increase cannot be described by a linear trend. Fluctuations of marine faunal biodiversity, extinctions and recoveries have been correlated by many authors with changes in both the global arrangement of lithospheric plates and the size of habitable marine areas. Both the diversity and the area of shallow-marine seas are known to have decreased during the Permian due to the formation of the supercontinent Pangaea, although the end-Permian mass extinction coincided with a transgression (Hallam and Wignall, 1997, 1999). By contrast, increasing diversity and endemism in the Mesozoic is correlated with the progressive opening of the Tethyan and Atlantic oceans, which resulted in greatly increased areas of shallow seas.

The temporal coincidence of changes in biodiversity and changes in the areas of habitable shallow marine waters appears to reflect a causal relationship, and gives impetus to palaeobiogeographic correlations based on the species–area relationship. MacArthur and Wilson's ETIB suggests that the same relationship between the equilibrium number of

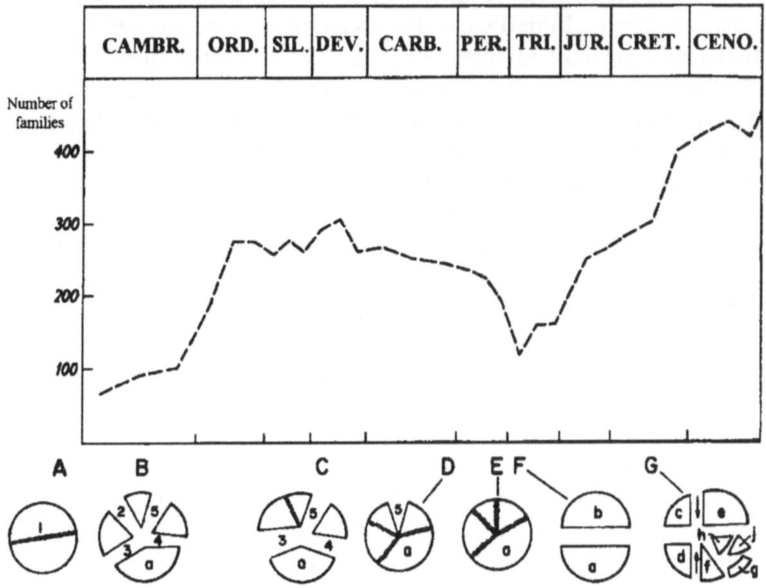

Figure 7.9 Diversity changes during the Phanerozoic and correlation with the history of the configurations of major lithospheric plates. A, Eocambrian suturing of the Pan-Africa–Baikalian (1) system and formation of Pangaea I (= Rodinia); B, fragmentation of Pangaea I with opening of pre-Caledonian–Acadian suture (2), pre-Appalachian (3), pre-Hercynian (4) and pre-Uralian (5) oceans in Cambro-Ordovician time; C, suturing of Caledonian–Acadian belt in Siluro-Devonian time; D, suturing of Appalachian–Hercynian belt in Late Caroniferous–Permian time; E, formation of Pangaea II by suturing of the Urals in Permo-Triassic time; F, opening of Tethys in Triassic–Jurassic time; G, a progressive fragmentation of Gondwana from the Cretaceous to Recent, opening of the Atlantic and closing of Tethys by alpino-himalayan suturing. Continents: a, Gondwana; b, Laurasia; c, North America, d, South America; e, Europe; F, Africa; g, Antarctica; h, India; j, Australia.

Source: Valentine & Moores (1970). Modified with permission of the authors and *Nature*.

species and the size of an island can be applied to regions of continental size. These authors have demonstrated that in Recent environments the species–area relationship is quantitatively predictable and best described by Preston's (1962) equation:

$$S = kA^z \tag{5}$$

where S is the number of species, A is the area and k and z are estimable parameters. In particular, k is an empirically determined constant taken into account for environmental features other than area controlling species diversity. It represents the biotic richness of an area depending on the taxon and biogeographic region. For example, modern tropical communities can contain more than an order of magnitude more species than modern Arctic communities within an equal area, and k would tend to normalize for this difference (Valentine and Jablonski, 1991). The parameter z accounts for differences between taxa. Species–area relationships may be also represented as log–log plots, therefore equation (5) can be expressed as:

$$\log S = z \log A + \log k \tag{6}$$

In this equation z describes the slope of the log–log relationship and $\log k$ describes its intercept. According to MacArthur and Wilson (1967), values of z fall in most cases between 0.20 and 0.35 for islands, and between 0.12 and 0.17 for non-isolated sample areas in continents (or within large islands), thus the slope of the log–log plot is steeper for islands. Any reduction in island area therefore lowers diversity more than a similar reduction of sample area from a contiguous mainland habitat.

Fig. 7.10A shows how diversity changes if habitable area is changed. Flessa (1975) demonstrated that z values are close to 0.26 for species and vary between 0.2 and 0.4 for species, genera and families. Starting with an area containing 100 species, a ten-fold increase in area can lead to an increase of biodiversity by about a factor of 2. In shallow-marine environments, this result can be obtained with eustatic sea-level rises, although the effects of sea-level changes, as shown below, may deviate from predictions based solely on equation (5).

The question of the extent to which the increase in diversity during the Phanerozoic may be accounted for simply by changing the extent of endemism has been addressed by Schopf (1979). Consider the curve of Fig. 7.10B with A (or X in Fig. 7.10A) = 10 and z = 0.40; the resulting value of S is 400. Now, cut the original area into two smaller areas (for example, growth of an isthmus cutting a continuous marine area into two distinct seas), the size of each area being one-half of the original area. The resulting number of species of each area is reduced by one-half to the z power: for $A = 5$ and $z = 0.40$ the resulting value of S is 255 for each area. It is important to note, however, the increased total diversity, from 400 to 510 species, i.e. the sum of the species contained in each of the two areas. We can move in the opposite direction, joining two areas each containing 255 species, or 510 total

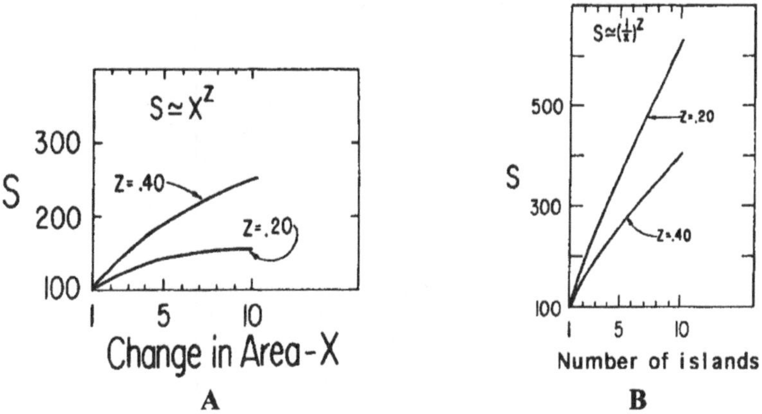

Figure 7.10 A: two curves of change in number of species (*S*) against increasing habitable area (*X*) with *z* values of 0.20 and 0.40. *B*: as a single area is subdivided into 10 smaller areas (or islands, or provinces), each 1/10 the size of the original, the number of species on each area is proportional to $(1/X)^z$. The curves show, for the two *z* values, the total number of species from *X* number of areas.

Source: Schopf (1979). Reproduced with Permission of Oregon State University Press.

species, in a single area twice the size of a single original area: only 400 species can exist in this single area. According to MacArthur and Wilson (1967), this implies that a significant extinction must occur to maintain the equilibrium. It is also possible to read Fig. 7.10B by subdividing the total area ($A = 10$) into ten areas, each considered as a single province. For a ten-fold reduction in area, with each province one-tenth the size of the original, the total fauna (i.e. the sum of the numbers of species of each area) becomes four to six for *z* values of 0.40 and 0.20 respectively.

The area effect is not the sole controlling factor. Changes in endemism increase faunal diversity, which changes in parallel. As Fig. 7.11 shows, changes in diversity driven by endemism (measured as the number of faunal provinces involved) are two to three times those that simply reflect an increase in the areas of available habitats. According to Schopf, significant diversity changes predominantly reflect changes in endemism more than the size of habitable marine areas.

Wyatt (1993) reached similar conclusions. He plotted shallow-water marine diversity for the Phanerozoic on the basis of a model that takes into account the shapes of continental blocks and the species–area relationship predicted by equation (5). He calculated the effects on

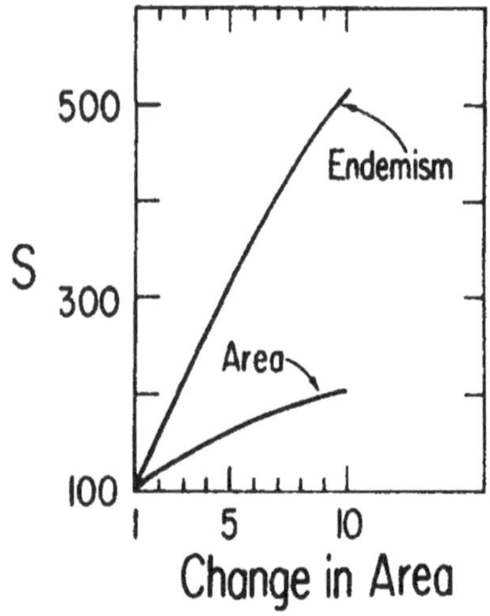

Figure 7.11 Changes in total species diversity (*S*) as the effect of change in area in degree of endemism (representing the number of areas, islands or faunal provinces). The curves are mean values for the range of values shown in Fig. 7.10.

Source: Schopf (1979). Reproduced with Permission of Oregon State University Press.

shallow-water areas of changing sea level, assuming that the continental shapes of the past followed the same patterns in terms of area–height relationships (hypsometry) as those of the present. This enabled him to draw diversity curves for different levels of endemism for each continental block. The results (Wyatt, 1993) show that the degree of endemism through time will have an important effect on any diversity curve. The magnitude of such an effect could be much greater than the changes due to changes in sea level, or changes in the number of faunal provinces. Wyatt's curves show a number of decreases in diversity. Some of these roughly correspond to times when the numbers of faunal provinces also decreased (the end Ordovician, end Devonian and end Permian mass extinction events), but others show no corresponding decrease (Middle Cambrian, end Cretaceous and end Eocene). In particular, both the end Cretaceous and the end Eocene mass extinctions occur at times of increase in the number of provinces. However, Wyatt's estimates did not take into account very detailed pre-Triassic reconstructions now available that show large numbers of small blocks that eventually coalesced to form Eurasia. These would undoubtedly have increased diversity.

7.4.2 Biogeographic response to sea-level changes

Sea-level fluctuations lead to variations in the size of habitable marine areas that could be described in terms of the species–area relationship. This may provide a useful tool allowing both diversity trends and levels of endemism to be predicted. As indicated, the increased size of an area resulting from a sea-level rise may accommodate a larger number of species (as predicted by the species–area curve). It may also lead to a reduction in the total diversity by the decrease in endemism that follows as formerly separated provinces merge. This is the case in the early Late Cenomanian transgression and is demonstrated by the turnover rates of ammonites (Monnet and Bucher, 1999). Conversely, however, falls in sea level may restrict the range of habitats within areas inhabited by a community or reduce the number of species as a result of the reduction of habitable area.

A fall in sea level does not always imply a reduction in diversity. If barriers emerge and fragment a previous shallow sea province, speciation can be enhanced and diversity may actually increase. Fig. 7.12 reproduces a model proposed by Valentine and Jablonski (1983b): a sea-level drop from coastline (3) to coastline (2) will increase provinciality while regression from coastline (2) to coastline (1) will reduce it. However, a sea-level rise does not always result in a reduction of endemism. In an epicontinental sea, transgression or regression may increase provincialism, depending on the original geographical (morphological) conditions.

Species–area predictions must take into account the hypsographic curve, as demonstrated by Wyatt (1987, 1993, 1995). Because the response of a continental block to sea-level change depends on the size of the block itself, the global diversity curve does not vary in a straightforward manner with respect to sea level. Wyatt's curve, published in 1987, masks the changes in individual blocks because it shows little change at the end of the Cretaceous where the theoretical loss in diversity, calculated by equation (5), for Laurasia alone is of the order of 25%. Wyatt (1987, p. 119) suggested that the Mesozoic/Cenozoic change 'may be in part a "Laurasia effect", and this may have coloured perceptions of the magnitude of the change elsewhere'. By contrast, the end Permian extinction shows a global reduction of both shallow sea area and diversity.

There is no general agreement on methods of discriminating the effects of area from those of environmental heterogeneity, as this inevitably increases with the expansion of the environmental area and decreases as the area reduces. Similarly, there is no simple relationship

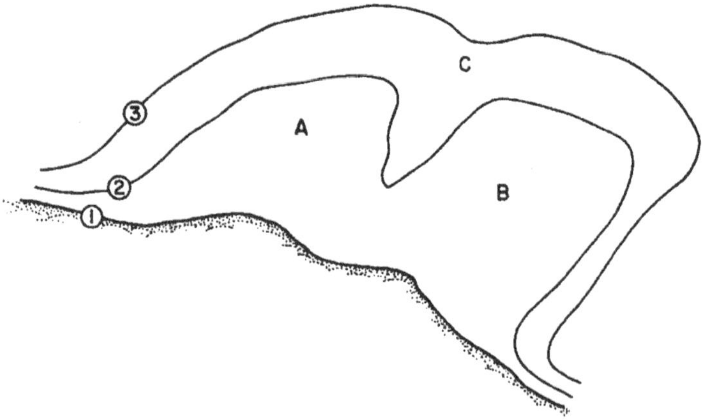

Figure 7.12 Hypothetical topography to show the possible effects of sea-level changes on provincialism in an epicontinental sea. Low sea level results in a fairly straight coastline (1) embracing a single province. A transgression displaces the coastline (2), which is now broken by a promontory into two embayments A and B that may contain endemic species, identifying distinct provinces. Further transgression drowns the promontory and results in a reduction to a single province, C, in a large embayment delimited by coastline (3). Regression from the latter position may reproduce two distinct provinces if there is a sea stand near coastline (2), but regression from coastline (2) to coastline (1) will reduce provinciality.

Source: Valentine & Jablonski (1983b). Reproduced with permission of Academic Press.

between endemism and diversity as predicted by the species–area curve and the biotic response to sea-level changes. Some examples illustrate these problems.

Hallam (1977) has shown a complex correlation between the incidence of endemism and the degree of inundation of continents during the Jurassic. Fig. 7.13 shows the high endemism at generic level among European and East Asian bivalve genera at the beginning of the Jurassic, when the percentage of continents covered by the sea was low. Endemism declines during subsequent sea-level rises, reaching a minimum in the Middle Jurassic. Interestingly, diversity, expressed as the number of bivalve genera, peaks at the same level. From the Bathonian, endemism increases again, with a new sea-level rise, reaching a peak in the Tithonian when sea level falls abruptly. Apart from a small drop in the Callovian, bivalve diversity does not change significantly after the Middle Jurassic peak.

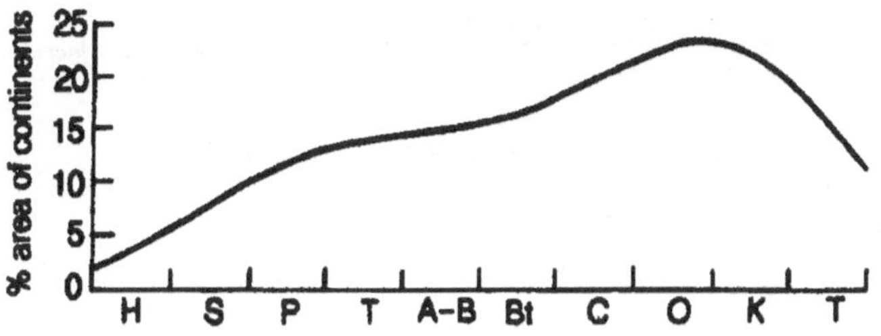

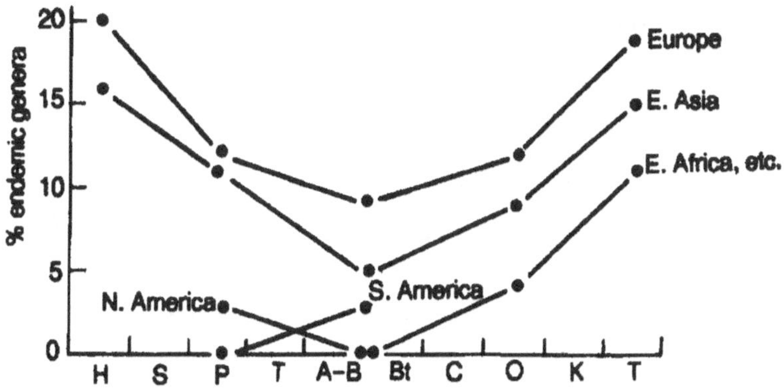

Figure 7.13 Above: the percentage of continents covered by sea during the Jurassic period. Below: the temporal variation in endemism of Jurassic bivalve genera for different regions.

Source: Hallam (1977). Reproduced with permission of the Paleontological Society.

The biotic component cannot be disregarded, and different taxa respond differently to sea-level fluctuations. During sea-level rises endemism in ammonites apparently falls (House, 1985), but the complex pattern of sea-level changes and variations in the proportion of endemism shown by Enay (1980b) in the Jurassic suggests that such a simple relationship cannot be generalized. The magnitude of the sea-level fluctuation and the duration of the low sea-level stand are also important.

Fig. 7.14 presents three hypothetical interactions between sea-level changes and changes in biodiversity and endemism in benthic and nectobenthic faunas on an epicontinental platform. The magnitude of the sea-level change is assumed to be important. Let us equate these different assemblages with distinct provinces. In periods of high sea

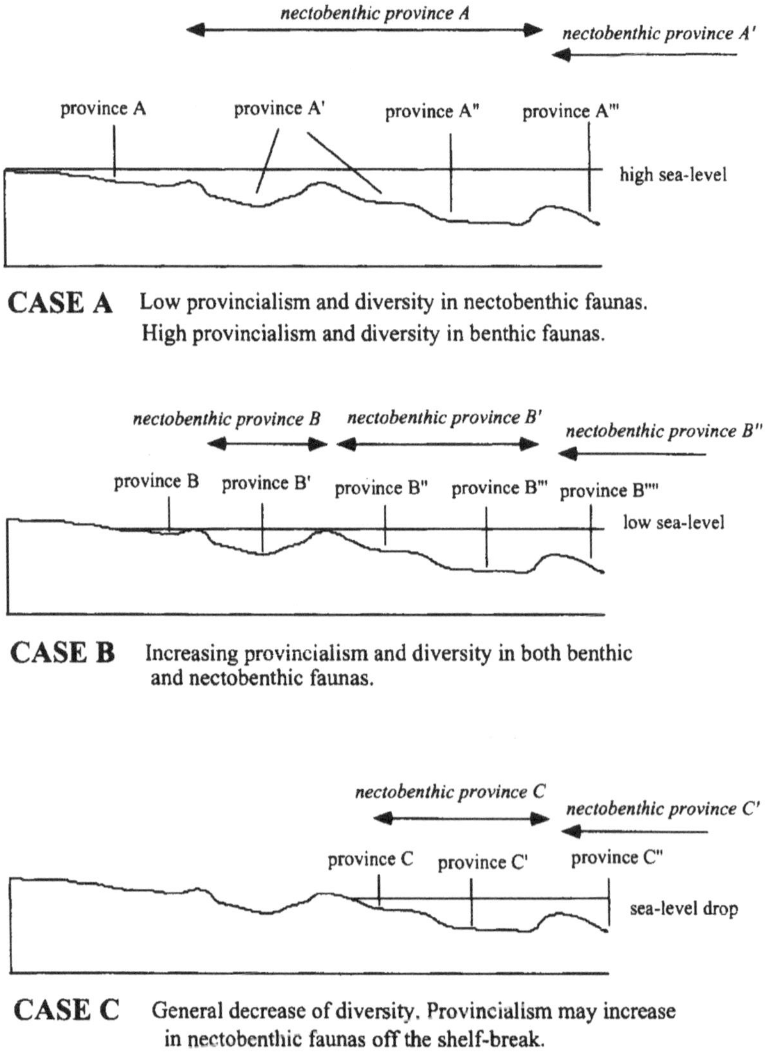

Figure 7.14 Hypothetical relationships between sea-level changes and provincialism for benthic and nectobenthic faunas on an epicontinental platform. See text for explanation.

level (case A) distinct assemblages of nectobenthic faunas, for example ammonites, are assumed to live on the shelf and off the shelf-break. In contrast, benthic assemblages show a clear differentiation. If sea level falls (case B), provincialism may increase, giving rise to distinct assemblages. The habitable area of benthic organisms shrinks, leading to stronger selective pressures that enhance speciation. Provincialism also increases in nectobenthic faunas because of increasing habitat heterogeneity. Among possible biotic responses are the synecological relationships between the nectobenthic organisms feeding on the benthos (others may feed on plankton), new trophic opportunities offered by a renewal of benthic faunas, and a new trophic web within the benthos itself. These relationships involve trophism-related adaptive responses and selective pressures. In distal, i.e. off shelf-break, epioceanic environments speciation factors in nectobenthic organisms such as ammonites seem to have been related to fluctuations in ocean productivity rather than wholly or partly to sea-level change (Cecca, 1998). According to Schopf's (1979) models, overall diversity should increase in case B. However, if a major sea-level fall occurs (case C), both number of provinces and diversity decrease for the benthos because wide areas of the platform are emergent. Nectobenthic habitats shrink, giving rise to strong selective pressure in more distal habitats. This situation has been described by Hoedemaeker (1995) and used to explain important faunal turnovers in Early Cretaceous ammonites. As in Fig. 7.12, these developments may equally be read in the opposite direction, from case C to case A. From B to A the transgression reduces both provincialism and diversity.

Aberhan and Fürsich (1997) reached similar conclusions regarding the importance of sea-level change for the mass extinction at the Pliensbachian–Toarcian boundary. The model proposed, devised for bivalves, involves factors such as climatic change, variations in nutrient supply, relative anoxia, sea-level change, and ocean circulation. The significance of each of the other factors changes with sea-level change, but the model differentiates between early transgression, when marine species diversity increases, and peak transgression, during which diversity decreases. Following peak transgression species diversity increases as sea level begins to fall (as in case B of Fig. 7.14). Because of the progressive isolation, the trend in reducing species diversity is reversed (as in case C of Fig. 7.14). This model explains why the Pliensbachian–Toarcian extinction was largely restricted to endemic bivalves.

7.5 Are extinctions related to sea-level falls? The biogeography of mass extinctions

Most episodes of faunal extinction coincide with important sea-level falls that reduced the extent of seas on epicontinental platforms. However, despite the fact that many authors invoke marine regressions as a cause of mass extinctions, there is increasing evidence to suggest that they are a result of other concurrent phenomena (Aberhan and Fürsich, 1997; Hallam and Wignall, 1997; Lethiers, 1998). Some faunal extinctions do not reflect sea-level falls and may even coincide with rises in sea level (Hallam and Wignall, 1999). For example, Ivany *et al.* (2000) suggested cooler winters as a possible cause of mass extinctions at the Eocene/Oligocene boundary. Nevertheless, in terms of the species–area relationship, the role of eustatic factors in the extinction of marine groups deserves to be developed further. MacArthur and Wilson's theory (1967) suggests that a reduction in the size of the area leads to a reduction of the number of species, according to equation (5). A fall in sea level implies a reduction of the marine habitable area but does not necessarily lead to extinction. The examples discussed above show that the species–area relationship is not a simple one. Thus, the biogeographic responses to sea-level falls and their possible implications for some cases of biogeographical selectivity in faunal extinctions are of particular interest.

Crowding and competititon caused by reduction of habitat size might be expected to be major agents of extinction, but we have seen that these factors may lead to speciation and increase diversity. Extinction may result from the contraction of individual populations to sizes that are no longer viable, as a result of the dramatic reduction or disappearance of habitats. The amplitude of the sea-level fall may be important (Aberhan and Fürsich, 1997), but no mass extinction resulted from the sea-level falls of Pleistocene glacial phases that exposed large areas of continental shelves and dramatically reduced areas of shallow waters. The duration of low sea-level stands has been suggested as relevant to the dynamics of mass extinctions, but most taxa are resistant to even prolonged low stands (Jablonski, 1985). Nevertheless, marine regressions may lead to extinction through the elimination of habitat types, and hence the reduction in environmental heterogeneity (or habitat diversity) at the community level. Sea-level change *per se* is not the sole cause of either the extinction of endemic populations or the disruption of biotic provinces. There is no common cause for mass extinctions, as they are due to a complex interaction of a variety of parameters.

7.5.1 Biogeographical accommodation during sea-level falls

The examples illustrated below indicate the relative roles of biogeographical factors that cannot be simply correlated with the species–area effect.

The well-known mollusc assemblages of areas of the eastern Pacific outside the Tropics provide a test for the biogeographic effects of Pleistocene sea-level variations on benthic faunas (Valentine and Jablonski, 1991, and references therein). The bioprovinces defined in recent faunas can all be recognized in the Pleistocene together with all living communities and associations. Environmental continuity was clearly maintained during glacial low stands because communities that would have been severely affected, such as those on rocky shores, can be traced as species-rich faunas in Pleistocene high-stand deposits and are equally rich today. The effect of decreasing habitat area has been calculated by Valentine and Jablonski (1991), using Preston's species–area equation (5). To do this, it is necessary to make assumptions as to the shelf width during sea-level falls based on the reconstruction of the shelf slope and on the position at which the thermocline intercepts the sea-floor. The thermocline is considered a barrier for benthic organisms because it determines the outer margin of the shelf environment. Taking into account the fact that the Californian shelf is narrower than the average and terminates at a much shallower depth, and assuming a depth of 110 m for the shelf-break and sea-level falls of between 60 and 150 m, the shelf area would be reduced to between ~50% and <10% of its present area. The species–area equation (with z values varying between 0.20 and 0.26) predicts species losses that range from 13–16% to 24–30% for shelf area reductions of 50% and 75% respectively. However, local extinctions were minimal, and fall well below the predicted values.

Thus, the apparent lack of response of the shelf fauna to important sea-level changes remains an open question. For the Pleistocene Californian molluscs, it is impossible to rule out the possibility that the effects of sea-level variations have gradually accumulated since the early Neogene, making the brief intervals between glacial eustatic falls insufficient for faunal replenishment. In fact there is no evidence of any replacement species and the present fauna could already be adjusted to low sea levels. It may be, however, that, the low stands were too short to produce all of the extinction predicted by the species–area equation. The reduction of the shelf area caused by sea-level falls may of course have led to local extinctions, to the disappearance of local populations of a species, but not to definitive extinction of all the populations that make up a species. Populations of the locally extinguished species survived, either in deeper water or to the south in warmer waters, and then returned during high stands, when the capacity of the wider shelf to accommodate them increased again and the temperature rose. Latitudinal migrations have been common and the lack of all the extinction required by the species–area relation can be explained by a biogeographical, rather than an evolutionary, response to sea-level falls.

Taviani (1998) suggested that the rate of sea-level change may have induced local extinctions in Pleistocene reef-associated molluscan assemblages in the Red Sea. Rapid sea-level rises (Collina-Girard, 1999) and falls would have impeded the setting of reasonably diverse coral reefs, as they provide insufficient time for full topographical expansion and the formation of exploitable habitats. Montaggioni (2000), focused mainly on examples from the IWP area, has shown that the rapid postglacial sea-level rise was accompanied by the settlement of three successive reef generations since 18,000 years ago, identifying two major reef drowning events, from 14,700 to 13,800 and from about 11,500 to 10,000 years ago. The present biota represents a recolonization which began 10,000 years ago. Taviani invokes a biogeographic accommodation of Red Sea taxa in the refuge area of the Indian Ocean.

The decrease in size of continental shelves that results from marine regressions is not in itself sufficient to explain mass extinctions. Hypotheses based on area effects in continental shelf benthos have not taken into account the taxonomic richness of the shallow-water biota around oceanic islands that act as refugia. The role of refugia has been evident in all extinction episodes. The 'Lazarus effect' (Jablonski, 1986a), which reflects the reappearance of apparently extinct taxa after a mass extinction event, has been explained by postulating that the refugia available to some organisms were effective for millions of years.

Jablonski (1985) observed that while small sea-level falls eliminate large areas of shallow epicontinental seas (that in some geological periods were considerably wider than today), the perimeters and shallow-water areas of conical oceanic islands slightly increase. As a result, the destruction of habitat types is limited in oceanic settings. Jablonski's study of marine benthic invertebrate families on Recent oceanic islands provides an estimate of the proportion of the marine biota that would not be affected by sea-level-induced reductions of shallow-marine habitable areas. The results show that a large proportion of today's shallow-water families are represented on oceanic islands, and suggest that reduction in continental shelf areas during marine regression cannot by itself explain mass extinctions. At most, only a 13% extinction of Recent families would be caused by a marine regression that eliminated all continental shelf families. The biological crisis at the Permo-Triassic transition, the most severe of the Phanerozoic, has been taken as an example of the relationship between a major sea-level fall and mass extinction. However, Jablonski's results show that the regression of epicontinental seas cannot explain the magnitude of such an extinction.

The lack of mass extinctions during the rapid glacio-eustatic oscillations of the Pleistocene may suggest that the explanation for the association of regressions and extinctions in the geological past may lie in the prolonged duration of sea-level falls. However, present distributional data do not support such an explanation because at least 78% of families are represented on two or more of the islands taken into account by Jablonski, indicating large populations distributed over many habitat patches and localities. Whereas individual islands may be subject to habitat alteration or random extinction, a rich and stable pool of taxa could persist even during prolonged regression, if all other factors remained constant.

Jablonski's (1986a) biogeographical explanation of the 'Lazarus effect' has been recently challenged by Wignall and Benton (1999). These authors suggest that the abundance of Lazarus taxa during their outage is probably a reflection of the extreme rarity of organisms in the time interval corresponding to the prolonged stressful conditions associated with the extinction crisis (Wignall and Benton's 'lag phase'). However, their alternative hypothesis was tested with examples of the postulated extinction among terrestrial vertebrates in the Late Carnian, for which refugia have not been invoked, and the end Permian extinction. For the latter, Wignall and Benton (1999) stress the rarity of refugee taxa in the Lower Triassic terranes of Japan and western North America. This contrasts with the high percentage of Recent skeletonized marine shallow-water families represented on oceanic islands. However, high percentage might be explained as due to the fact that Recent data do not reflect a 'lag phase', as no mass extinction took place in the Pleistocene. The hypotheses of Jablonski and of Wignall and Benton are not mutually exclusive, because during the 'lag phase' following a mass extinction like that at the end of the Permian, Lazarus taxa were scarce, even in their potential refugia.

7.5.2 Biogeographical structure and biogeographic selectivity

Jablonski (1985, 1986b) and Jablonski *et al.* (1985) stressed the importance of biogeographical selectivity during episodes of mass

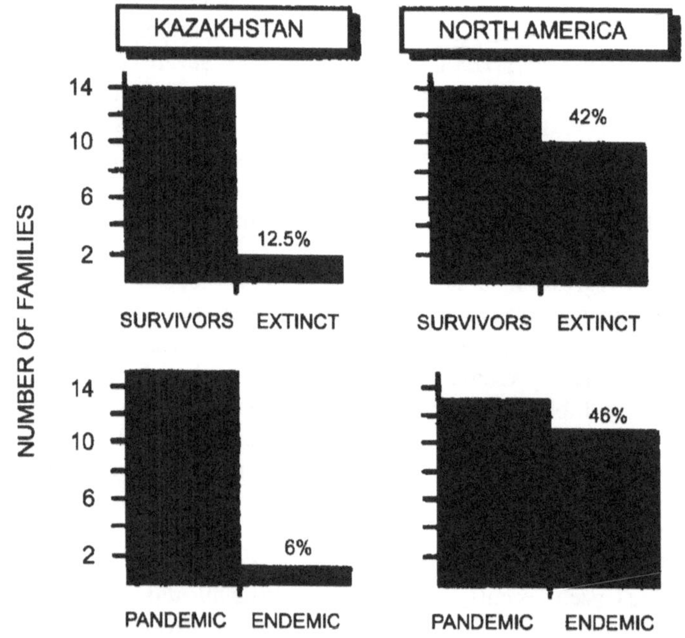

Figure 7.15 Patterns at the family level of the terminal Cambrian trilobite extinction in Kazakhstan and North America. The magnitude of the extinction in the two areas is compared above, and the proportion of endemic families *before* the onset of the extinctions is shown below.

Source: Westrop (1991). Reproduced with permission of the Paleontological Society.

extinction. Widespread taxa survive more than endemic forms, which are more abundant in tropical areas. It is to be expected, therefore, that higher extinction rates would be recorded in tropical areas because of the different biogeographical structure of the biotas in this peculiar latitudinal belt. Trilobite extinction patterns at the Cambrian–Ordovician boundary interval of Kazakhstan and North America (Fig. 7.15), studied by Westrop (1991), show that areas richer in endemic families experienced higher mortality. There is no relationship to latitude because both areas were in low latitudes at the time. Instead, the Kazakh succession corresponds to deep basin to slope environments and the higher proportion of pandemic families most likely reflects the tendency of deep-water Cambrian trilobite faunas to be geographically widespread. In contrast, high-latitude shelf trilobites, as those of the Baltic province, were exclusively pandemic, crossed the Cambrian–Ordovician boundary and did not experience extinction.

Westrop (1991) suggests (in agreement with Jablonski *et al.*, 1985) that latitudinal trends in the magnitude of extinction are influenced by differences in biogeographical structure and the selection of survivors is not reducible to properties of individual organisms.

It has been demonstrated that taxa characterized by planktotrophic larval development have wider biogeographical ranges than nonplanktotrophs. The latter are extinction-prone because they cannot maintain broad ranges to persist on islands because of the low probability of recolonization after local extinction. However, Valentine and Jablonski (1986) have shown that among prosobranch gastropod taxa with planktotrophic larval development survived the end-Cretaceous extinction as well as nonplanktotrophs. In contrast, Gallagher (1993) has shown that nonplanktotrophic molluscs, brachiopods, bryozoans, sponges and corals were less affected than planktotrophic forms during this mass extinction event. Jeffery (2001) has shown that there is no correlation between survivorship and larval strategy, size of geographic range, absolute geographic distribution of heart urchins at the Cretaceous–Tertiary boundary; on the contrary, feeding strategy is correlated with survivorship. Representatives of the most important groups in tropical benthic communities in shallow Permian seas, such as articulate brachiopods and stemmed crinoids, were probably characterized by a planktotrophic larval development (Strathmann, 1978; Valentine and Jablonski, 1983a). The diversification of post-Palaeozoic representatives has been constrained by the survival of taxa exclusively characterized by nonplanktotrophic larval development, which is commonly at a disadvantage in low-latitude habitats (Jablonski *et al.*, 1985). There are contrasting claims concerning the relative roles of planktotrophy and nonplanktotrophy in mass extinctions. Jablonski (1986b) observed that during normal periods of background extinction, planktotrophy ensures broad geographic range of constituent species and high species richness, enhancing survivorship at both specific and generic levels. These factors are ineffectual during mass extinctions, but broad geographic deployment of an entire lineage, regardless of the ranges of its constituent species, enhances survivorship.

This biogeographical factor seems to be more important than larval attributes to explain different selectivity regimes between background and mass extinctions. As far as mass (rather than local) extinctions are concerned, survivorship may not be reducible to the properties of individual organisms and a link between the biogeographical range and clade survival must be considered (Jablonski, 1985; Jablonski *et al.*, 1985; Westrop, 1991). The mass extinctions at the Cretaceous–Tertiary boundary seriously affected tropical faunas, but as far as bivalves were concerned, high extinction intensities occurred in shallow-water carbonate environments but not in other low-latitude settings (Jablonski, 1998). In conclusion, biogeographic structure and selectivity are general traits that are not comparable to the effects of sea-level falls in reducing the size of marine habitable areas.

7.6 Disturbance events and deviations from species–area predictions

Explosive volcanic eruptions, short-term climatic phenomena such as hurricanes, and even human impact are all part of what ecologists call **disturbance**. The study of recovery after disturbance events on islands (the most famous example is of the catastrophic eruptions of Krakatau, in the Sunda strait, which reduced the island to one-third of its original size and covered its biota with 60–80 m of ash) has important applications in biological conservation programmes. Similar data may be used to test MacArthur and Wilson's ETIB, which predicts community composition in conditions of substantial equilibrium.

Rosenzweig (1995) underlined the problems of scale in disturbance events, taking two extreme examples: an unusually powerful wave tearing away the residents of a patch of the rocky intertidal zone, and the asteroid impact responsible for the K-T mass extinction. The spatial scale is relevant to biogeography but in palaeobiogeography it must be coupled with the geological time-scale (Brown, 1988). If species and ecosystems have evolved within the context of disturbance regimes, as Whittaker's (1998) disturbed island ecology suggests, a perspective is offered to palaeontologists to explore the role of disturbance events in both microevolution and macroevolution (depending on the time-scale).

The biological response of present-day coral reefs and of marine invertebrates generally to disturbance has been intensively studied, but the time-scale (10 to 100 years) cannot easily be applied to fossil examples. The response of reefs to stress can be studied over relatively long time-scales, but obviously the phenomena responsible for disturbance events are different from those operating in the short term. Nevertheless, records of storm events in the sedimentary successions (tempestites) can be used to evaluate changes in fossil communities driven by disturbance. The time scale of Pleistocene glacial cycles ranges from 10^3 to 10^4 years and low temperature, falls in sea level and hydrological disturbances affected Pleistocene Red Sea coral reefs (Taviani, 1998). These disturbances resulted in local extinctions, but species were biogeographically accommodated in the refuge area of the Indian Ocean and were able to recover. This pattern is consistent with the relative stasis of coral reef communities over the past 10^5 years shown by Jackson *et al.* (1996), who invoke latitudinal compression of biogeographical ranges of tropical species and local extinction of metapopulations and metacommunities in response to glacial cycles.

There is no doubt that fossil communities provide significant information on long-term disturbance. Taviani (1998) has questioned the stability of the Tropics because of the effects of the glacial cycles on biotas. However, this raises the question of whether or not mass extinctions should be considered in the disturbance perspective. Rosenzweig (1995) cautioned that disturbance–diversity patterns rely on small-scale disturbances and mass extinctions have nothing to do with them. The dynamics of these crucial macroevolutionary events is very complex and generalizations lead to numerous contradictions. Each of the factors invoked to explain mass extinctions can result in disturbance, but if this is not associated with other phenomena, it may be ineffective. For example, the Late Pliocene turnover in Caribbean molluscs and corals was probably enhanced by the onset of glacial conditions. However, the fall in Late Pliocene temperatures was less than other falls recorded in the past million years that did not result in mass extinctions (Jackson *et al.*, 1996). Thus, the same type of disturbance, a glacial perturbation acting at a time scale of 10^3–10^4 years, may produce different effects.

Jackson *et al.* (1996) recall that the equilibrium theory of island biogeography and related models have been successful in explaining most of the variance in community composition, where changes over ecological and evolutionary time have been considered gradual and continuous. It might be expected as a result of this that speciation by natural selection would be gradual, but, interestingly, Jackson *et al.* (1996, p. 110) suggest that over geological time this prediction is false and that 'the controversy over community membership and stability cannot be resolved without palaeontological data to provide the necessary time perspective and replication on appropriately large spatial scales'.

Short- and long-term disturbance in ecological palaeobiogeography should both be considered at the appropriate scale, taking into account the importance of metapopulation and biogeographic structure concepts. The different biotic responses to sea-level changes discussed in section 7.4.2 often deviate from species–area predictions and tend to confirm the suggestion of Jackson *et al.* (1996) for long-term, macroevolutionary events.

8 Palaeobiogeography and palaeogeographic reconstructions

Palaeogeographic reconstructions based on palaeobiogeographic data start with compilation of presence/absence data (Rosen's rationale 1; see Chapter 2) which may be treated with PAE methods (see Chapter 5). The treatment of these data may lead to the recognition, subjectively or through statistic methods ('phenetics'), of characteristic assemblages which virtually coincide with bioprovinces (see Chapter 4). The methods used for palaeogeographic reconstructions assume those concepts and processes (rationale 4) with practical value for reconstructing Earth history listed in Table 2.1. We have already discussed some applications of these methods to selected case histories.

This chapter aims to illustrate the application of palaeobiogeographical data to classical cases of palaeogeographic reconstruction, such as marine palaeocorridors and displaced terranes. The focus is on reconstructions based on the application of palaeobiochoria, with special attention to geoecological assemblages, 'assemblage methods' of Rosen (1988b). The use of assemblages has been strongly criticized and dismissed by cladistic biogeographers who question the objectivity of bioprovinces. However, four points should be stressed:

1 numerous reliable palaeogeographic reconstructions have been based on assemblage methods (or could have been successfully based on this method, as illustrated by the famous case of the *Lystrosaurus* in both South Africa and South America), so they cannot be disregarded
2 some biotic assemblages, such as those of carbonate platforms, represent a characteristic geoecological assemblage which allows the recognition of warm and shallow-water low-latitude environments through the whole Phanerozoic
3 endemic faunas give evidence for palaeogeographic reconstructions in old geological periods such as the Early Palaeozoic, for

which geological and geophysical constraints are controversial (McKerrow and Scotese, 1990) and the cladistic biogeographic approach is unrealistic (the percentage of extinct taxa is too high and the phyletic relationship with modern representatives is too distant in time)

4 a pragmatic strategy of investigation must be followed in most cases: although I personally believe that PAE methods have a stronger heuristic value (including Early Phanerozoic) than assemblage methods, it is often impossible to obtain significant taxonomic lists at different OPGUs to generate reliable PAE patterns.

This last point must be clarified, and I take the liberty of using an example from my personal experience. I have been involved for many years in research on Tethyan Jurassic and Cretaceous ammonites. Tethyan faunas are characterized by the significant relative abundance of Phylloceratina and Lytoceratina, i.e. taxa which give far lower biostratigraphical resolution than Ammonitina. Because of this, Phylloceratina and Lytoceratina have been systematically disregarded in many biostratigraphical studies and it is virtually impossible (with some exceptions) to obtain faunal lists of the complete ammonite assemblage occurring in a given locality. This loss of information is certainly a more serious problem than the taxonomic difficulties existing in some other groups of ammonites (Cecca, 1999) and definitely hampers the use of PAE.

Methods based on assemblages can be used with rigour, provided that assumptions are clearly acknowledged and circularities avoided, although in some cases this can be objectively difficult.

Because some assemblages show closer taxonomic composition to each other than to others, cluster hierarchies or relative biotic distances are expressed statistically. On the other hand, the quantification of endemicity, expressed in absolute totals or as percentages of the whole biota of an area, generates patterns which can be suggestive of relative degrees of isolation of different areas (rationale 2). However, the processes that led to isolation may be of different nature (geographical, ecological, etc.) and cannot be explained simply by patterns of endemicity. The latter give no indication of relationships between areas, thus limiting the significance of the quantification of the endemicity for palaeogeographic reconstructions. The inference of geological processes to explain isolation (rationale 3) depends on the knowledge of the relationship between geology and biology, which is generally poor. Therefore, biogeographical processes are often assumed (rationale 4).

The principles for using assemblages in palaeogeographic reconstructions have not been established and different authors have followed what Rosen (1988b) called 'simple rules' for interpreting assemblages. Concerning marine biotas, two cases can be distinguished:

1 sample localities share the same assemblage – in this case they occurred in the same or interconnected marine basins or oceans
2 sample localities have different assemblages – they occurred in different marine basins or oceans separated by either land masses or great oceanic distances.

Because it is practically impossible to find total difference or total similarity between assemblages, different levels of similarity are difficult to interpret. On the other hand, the ecophysiological characteristics of each group (for example the type of larval development) must be taken into account.

8.1 Geoecological assemblages

Patterns of endemicity derived from the analysis of geoecological assemblages provide indications of palaeogeographic discontinuity. Thus biogeographic data can generate palaeogeographic hypotheses which can be tested by independent lines of evidence from other disciplines. On the other hand, palaeogeographic hypotheses not based on biogeographic constraints are amenable to testing by biogeographic data. In this respect, specialists of Early Palaeozoic fossil groups are particularly involved in reconstructions of palaeopositions of major plates, whereas Mesozoic workers mainly deal with reconstructions of palaeopositions and displacements of the so-called terranes.

Cocks and Fortey (1982) and Fortey and Cocks (1986) proposed some principles which are valid also for more recent organisms, for the use of geoecological assemblages in problems of palaeogeographic reconstruction. Five criteria for recognition of past continental distributions from fossil evidence were listed by Cocks and Fortey (1982).

1 Faunal similarity alone is not sufficient to indicate geographic continuity. For example, not only pelagic groups but also deeper water benthic faunas can be similar across oceanic barriers, as these authors showed in the case of Iapetus and Tornquist oceans during the Ordovician. Inner shelf faunas from equivalent habitats must be compared to recognize provincial differences related to the past distribution of continental masses. The underlying

assumption is that habitats and ecology of the organisms concerned are sufficiently known.

2 Recognition of facies belts parallel to the edges of former continents can be used in conjunction with palaeobiogeographic distribution of inner shelf faunas, assuming that sediment types are also comparable.

3 Planktonic or epipelagic organisms are related to palaeolatitude. The faunas distributed around the palaeoequator (which can be delimited on the basis of distinctive organisms such as corals or other reefal elements) will have few species in common with high latitudes, whereas planktonic forms from intermediate latitudes will have some restricted species and some species in common with both higher and lower latitudes.

4 Duration of the larval stage, characteristics of larval dispersion and life habits of benthic organisms are also important, as discussed in Chapter 6.

5 Because of their mixed characteristics compared to faunas from neighbouring continents, faunas from terranes (indicated as 'volcanic islands' by Cocks and Fortey, 1982, p. 466) do not have primary importance in the reconstruction of the past distribution of continental masses. Nevertheless, they may be important as potential staging posts.

Cocks and Fortey also proposed a model to illustrate the application of these criteria while cautioning that it does not allow the recognition of both transcurrent movements within the same latitudinal belt and effects of eustatic variations.

In Fig. 8.1A faunas of shelves separated by an oceanic basin are the same but if ocean width increases, endemic forms may appear among the less mobile taxa (the ocean would be a filter for benthic faunas); planktonic faunas are the same. In Fig. 8.1B, benthic faunas of separated shelves show differences (note that faunas of a single N–S oriented shelf, such as modern molluscs along the North American shelf, also exhibit latitudinal variations); planktonic faunas will reflect latitude. However, deep-water benthic faunas within a single ocean may not be as uniform as Cocks and Fortey suggested (Rosen, 1988b). The Early Ordovician cyclopygid trilobites, which are regarded as having lived at approximately 200–500 m water depth, seem to have been restricted to the Gondwana margin whereas graptolites are pandemic (Fortey *et al.*, 1989; Cooper *et al.*, 1991).

A 'classical' example of use of the distribution of geoecological assemblages can be drawn from Fortey and Cocks (1986). The authors

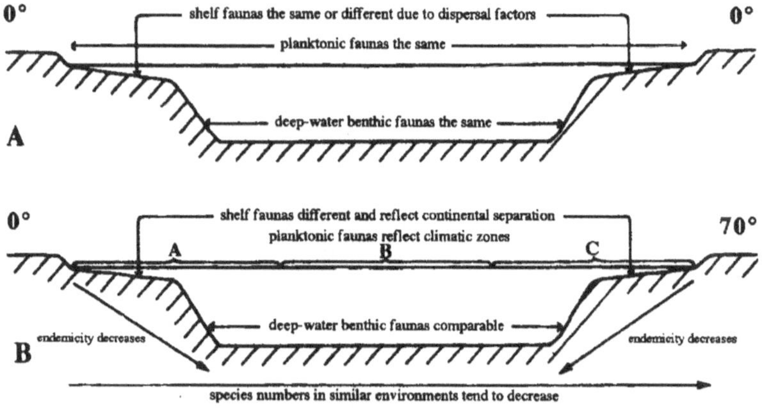

Figure 8.1 Comparison of faunas on opposite sides of an ocean: A, at the same latitude; B, at different latitudes and at the same longitude.

Source: Cocks and Fortey (1982). Reproduced with permission of the authors and the Geological Society.

focused on the distribution of marginal biofacies to recognize past continental edges. According to the model of Cocks and Fortey (1982), these biofacies tend to have widespread taxa and are of little use to infer palaeolatitude. However, they can indicate the proximity of former ocean basins even where field evidence for this may not be decisive. An example of the distribution of the isograptid and other graptolite biofacies in the Early Ordovician is illustrated in Fig. 8.2. The isograptid facies indicates proximity of ocean basins whereas the non-isograptid facies characterize epicontinental seas. The isograptid biofacies is seen to be located at the edges of former continents, though the reconstruction is derived from a synthesis of numerous lines of evidence.

The Early Ordovician distribution of graptolite assemblages has been proved to be consistent with that of trilobites; inshore-offshore gradients have been recognized too (Cooper *et al.*, 1991).

8.1.1 Limitations of the biogeographic methods: some controversial cases

The resolving power of biogeographic data in problems of palaeogeographic reconstruction is limited. This may be a function of limitations of a method based on the use of geoecological assemblages and on the subjectivity inherent in qualitative definitions of the assemblages themselves. Although all methods of biogeographic analysis derive patterns which must be tested with independent lines of evidence from other

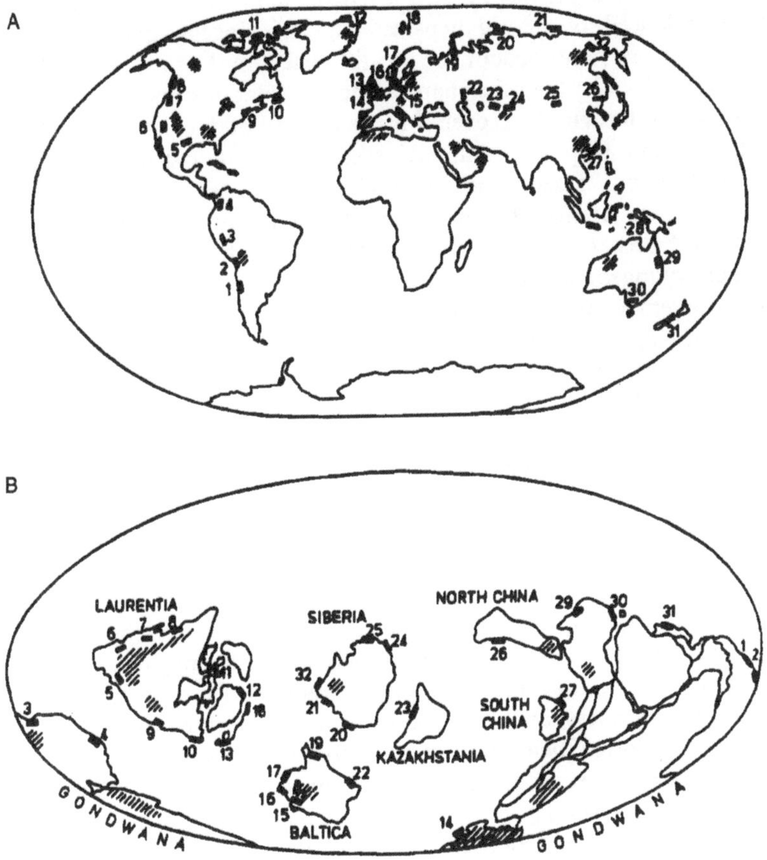

Figure 8.2 (A) Modern locations of Early Ordovician isograptid facies (black rectangles), indicative of the proximity of former ocean basins, and non-isograptid graptolite faunas (diagonal shading), indicative of shelf areas; (B) distribution of isograptid and other graptolite facies on a Lower Ordovician palaeogeographic reconstruction.

Source: Fortey and Cocks (1986). Reproduced with permission of the authors and the Geological Society.

disciplines (and vice versa), in some cases biogeographically constrained results are not conclusive or do not stand the test.

There is no doubt that pre-Mesozoic reconstructions of the palaeogeographic positions of the major plates are far less constrained than Mesozoic–Cenozoic ones because no remnants of oceanic continental crust older than the Middle Jurassic are preserved. Reconstructions for the different periods of the Palaeozoic are still controversial and the unbalanced role played by palaeomagnetic, biogeographic and geologic evidence may account for the differences in the proposed palaeogeographies. On the other hand, different solutions may often be at variance with each other, even if the same kind of data has been taken into account with the same weighting. Different palaeogeographic reconstructions based almost or exclusively on, say, palaeomagnetic data may result from either the different interpretation of the same data or discrepancies between data obtained by different research teams.

Palaeobiogeographically constrained reconstructions also give rise to divergent solutions (Jablonski *et al.*, 1985), thus highlighting limitations of biogeography in resolving palaeogeographic problems. One of the most controversial cases is the so-called Boucot and Gray's (1983) 'Paleozoic Pangaea'. These authors have spectacularly shown the disagreement between different competing palaeogeographies focusing on the Devonian by comparing different reconstructions for the same geological period. According to Boucot and Gray, who recognize the existence of Gondwana in the southern hemisphere, a 'clustered' arrangement of Eurasia would better fit the biogeographic and geologic constraints (the latter corresponding to climate-diagnostic lithofacies such as evaporites, calcretes, and coals) than reconstructions with dispersed Eurasian terranes mainly based on palaeomagnetic data. It is worth noting that these two authors, admittedly, consider palaeomagnetic data less reliable in the Palaeozoic than other data. Boucot and Gray plotted their biogeographic subdivisions and the other kinds of data on palaeogeographic maps proposed by other authors and analysed the resulting inconsistencies. It is particularly impressive to see that their palaeobiogeographic subdivisions do not fit the palaeogeographic scheme of Bambach *et al.* (1980), even though the same or similar biogeographic data were used. Such data are mostly based on distributions of shelf organisms which are useful in recognizing provincial differences related to the past distribution of continental masses. The biogeographic subdivisions of Boucot and Gray have been revised by other authors and definitely do not fit the more recent, palaeomagnetically better constrained, reconstructions of McKerrow and Scotese (1990) (Hallam, 1994).

Another case of striking contrast in palaeobiogeographic reconstructions is that of the Tornquist Sea, discussed in Chapter 5. The two different hypotheses proposed for the Early Ordovician palaeogeography by Cocks and Fortey (1982) and Paris and Robardet (1990) are both biogeographically constrained and virtually based on the same data set. In fact, despite the assertions of Fortey and Mellish (1992) and

Hallam (1994), the reconstruction of Paris and Robardet is not based on pelagic groups, such as acritarchs and chitinozoans. These organisms have been taken into account for Silurian and Devonian reconstructions (Paris and Robardet, 1990, pp. 202–203, 206) together with trilobites, brachiopods, graptolites and ostracods, but for the Ordovician (p. 200) the authors quoted some chitinozoans, while acknowledging their uniform distribution, and did not constrain their reconstruction on the basis of the latter organisms. Having carefully read and critically compared the two papers, I feel that the differences between the reconstructions of the Early Ordovician palaeogeography proposed by Cocks and Fortey and Paris and Robardet are simply the result of different interpretations of the same database. As already mentioned, the hypothesis of the Tornquist Sea has been subsequently supported by both PAE and cluster analysis of fossil asemblages, and by a survey of palaeomagnetic data independently. Further support has been given by the recent palaeoceanographical model proposed by Christiansen and Stouge (1999) for the Arenig. These authors used the shifts in Hadley circulation caused by orbital variations and the rotation rate of the Earth (which was spinning about 12% faster than today) to model surface current flows which are consistent with the palaeobiogeographical patterns and with the presence of the Tornquist Sea.

These controversial cases, which are not unique, do not weaken the biogeographic method in constraining palaeogeographic reconstructions but simply show its limitations. As Smith (1988) has shown in his biogeographically constrained reconstruction of the tectonic history of Asia (see Chapter 5), the most rigorous approach in the study of a palaeogeographic problem is the use of different analytical treatments of the biogeographical database. I think that palaeogeographic conclusions based on geoecological assemblages, but accompanied by other analytic treatments (BPA, PAE, methods of multivariate analysis, etc.), have the best chance of standing testing by independent lines of evidence.

8.2 Marine corridors and routes

Most palaeobiogeographic studies of marine fossil faunas are aimed at seeking dispersal routes permitting the exchange of species belonging to different biotas.

Some marine corridors have been inferred from geological and palaeontological records and correspond to real, past, physical routes permitting dispersion (e.g. epicontinental seaways separating two land areas), a process which tends to maintain the gene flow if the corridor

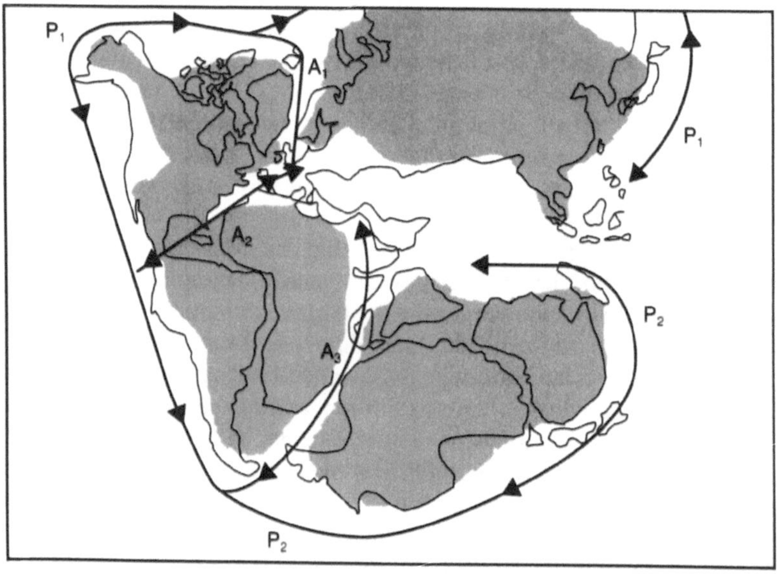

Figure 8.3 The marine corridors (A_1, A_2, A_3) across Pangaea and routes (P_1, P_2) around its margins during the Jurassic. Land areas are represented by a stippled pattern.

Source: Hallam (1983). Reproduced with permission of Elsevier Science.

runs along the same latitudinal belt. Conversely, longitudinal corridors may cross climatic barriers and act as filters.

The recognition of marine corridors in old geological periods, particularly in the Early Palaeozoic, may be problematic (e.g. the problem of the Tornquist Sea). In geologically younger examples it is the age of their opening that is mainly questioned.

Hallam (1994) has summarized the most important corridors which were established across Pangaea as the supercontinent began to break up and sea level rose, as well as the peripheral boreal and austral routes.

8.2.1 Peri-Pangaea routes

Faunal movements have been proved to have occurred along the shelves bordering Pangaea. A Boreal route (P_1 in Fig. 8.3) is proved by unambiguous distribution patterns during the Jurassic: for example, the distribution of the Pliensbachian ammonite family Amaltheidae in north-western Europe, East Greenland, westermost Canada and USA, Siberia, and the distribution in more or less the same areas of Middle–Late Jurassic bivalves (*Buchia*) and ammonites (Enay, 1972, 1985; Callomon, 1985; Smith and Tipper, 1986; Zakharov *et al.*, 1996). An Austral route (P_2 in Fig. 8.3) is suggested by the distribution of Jurassic bivalves (Damborenea, 1993; Gardner and Campbell, 1997).

8.2.2 Trans-Pangaea corridors

The 'Viking Corridor' (Westermann, 1993) corresponds to a Jurassic epicontinental seaway, which roughly stretched along the rifted structure of the future northern branch of the Atlantic Ocean (A_1 in Fig. 8.3). Geological data do not support a continuous seaway before the Toarcian (Bassoullet *et al.*, 1993), although faunal distributions of ammonites and bivalves suggest it had been open since the Pliensbachian (Hallam, 1977; Enay, 1985). The effectiveness of the 'Viking Corridor' for faunal exchanges during the Jurassic was affected by tectonic movements and eustatic changes (Enay, 1980b; Enay and Mangold, 1982; Westermann, 1993; Hallam, 1994) which controlled expansions and contractions of Boreal and Tethyan faunas.

The corridor linking East Africa with the southern Andes via Antarctica (A_3 in Fig. 8.3) was probably not established before the end of the Jurassic. Although a proto-Mozambique Channel had been open since Callovian (Enay *et al.*, 1993), there is no geological evidence supporting a direct link between East Africa and the Andes. Previously quoted biogeographic arguments were based on palaeontological misidentifications (Riccardi, 1991; Hallam, 1994). On the other hand, ammonites, bivalves and belemnites support a link being completely established since Tithonian (Enay, 1972; Mutterlose, 1986; Hallam *et al.*, 1986; Riccardi, 1991).

The next corridor (A_2 in Fig. 8.3) deserves a special section (Box 8.1) because the age of its opening has stimulated a strong, controversial debate between specialists.

8.3 The problem of the terranes

The Pacific sides of both Asia and the Americas are geologically characterized by assemblages of numerous microplates, or **terranes**, separated from the cratons by evident geological discontinuities. The terranes are identified on the basis of an internally consistent stratigraphy which is unique to each terrane. They are often designated as 'suspect' terranes because of uncertainty about their palaeogeographical

Box 8.1 The age of the opening of the 'Hispanic Corridor'

After the formation of the Pangaea supercontinent, the direct marine connections between Panthalassa and Tethys were interrupted. At the beginning of the Mesozoic, the western limit of the Tethys coincided with the present-day Mediterranean area. The rifting that led to the breakup of Pangaea and to the opening of this branch of the Atlantic, with the resulting separation of Laurasia and Gondwana, began in the Middle Triassic (Marcoux *et al.*, 1993). The communications were certainly re-established in the Callovian, when the formation of the oceanic crust of the Central Atlantic began. However, marine communications have been invoked prior to Callovian, as early as Early Jurassic, on the basis of similarity of ammonite and bivalve faunas from Europe and the American Cordillera (Hallam, 1983). Because some Tethyan taxa were found in allochthonous terranes, these affinities have also been explained by tectonic migration from West Pacific areas (Stevens, 1983; Ross and Ross, 1985) of terranes reached by larval dispersion. The crossing of Panthalassa would be facilitated by islands and atolls, subsequently accreted to the American craton, acting as staging posts (Tozer, 1982).

Two alternatives have been proposed to explain the faunal affinities between Western Tethys and the American Cordillera:

1 the existence of a non-oceanic trans-Pangaea seaway called the 'Hispanic Corridor' (Smith, 1983)
2 a model based on pantropical distribution which involves larval dispersion by west-to-east equatorial counter-currents (Newton, 1988).

The latter model will be discussed in detail in section 8.4. In this section we focus on the hypothesis of the 'Hispanic Corridor'.

The existence of this corridor is certain from Bathonian (Bartolini and Larson, 2001); its effectiveness is demonstrated from Oxfordian onward (Westermann, 1993). It has been postulated that this corridor could have intermittently connected, acting as a filter, Western Tethys and Panthalassa since the Pliensbachian (Smith, 1983; Smith and Tipper, 1986; Damborenea and Manceñido, 1979), or at least from Toarcian to Early Bajocian (Hallam, 1983). Plausibly, a shallow sea may have been formed as the result of

Box 8.1 continued

crustal thinning prior to the onset of the oceanic crust (before the Bathonian in this case). But what are the biogeographical arguments which support an earlier opening of the 'Hispanic Corridor'? Are these arguments tested by independent, geological, lines of evidence? A thorough, critical, discussion of facts and hypotheses is needed.

From the geological point of view no marine deposits (note that I am speaking of marine sediments and not of oceanic crust) older than Bathonian are known between West Morocco and Central America–Gulf of Mexico, either in land outcrops or in boreholes. In the Toarcian the westernmost Tethyan shores were located in Western Morocco (Bassoullet *et al.*, 1993). An Early Jurassic marine connection cannot be inferred on the basis of the geological data (Newton, 1990; Elmi, 1993; Dommergues, 1994; Cecca, 1999). The biogeographic arguments are mainly based on the distribution of both pectinoid bivalves (the genus *Weyla* has been frequently mentioned in the literature) and some ammonite genera. *Weyla* appears in the Late Hettangian of the Canadian Cordillera (Stikinia terrane) and during the Sinemurian it is recorded in both Southern and Northern American Cordilleras (Damborenea and Manceñido, 1979; Aberhan, 1998). This bivalve is recorded in the Pliensbachian of Morocco, France and Spain and in the Toarcian of Europe, Kenya, Madagascar, Saudi Arabia and Pakistan and eastern Pacific areas. This distribution has been thought to be suggestive of west-to-east faunal displacement from eastern Pacific areas to Western Tethys and eventually Eastern Tethys. This genus is lacking around southern margins of Gondwana and its earlier appearance in Western rather than Eastern Tethys has led authors to discard a southern Gondwanan route (P_2 in Fig. 8.3) for its expansion. Despite this, it has been considered a powerful argument for a Pliensbachian 'Hispanic Corridor'. Some points have to be stressed.

1 If the progressive time-migration pattern is considered a decisive one, the occurrence of *Weyla* in both France and Spain earlier than Morocco is inconsistent with its west-to-east range expansion. In fact, Damborenea and Manceñido (1979, p. 93) quote *Weyla* from middle Upper Pliensbachian marly limestones of northern Middle Atlas and from uppermost

Box 8.1 continued

Pliensbachian–Lower Toarcian strata of the eastern High Atlas, whereas it occurs in Lower Pliensbachian deposits of France and Spain.

2 Damborenea and Manceñido (1979, p. 86) stressed that 'facies seem to have played a decisive role in conditioning the last occurrence of representatives of this genus'. The Lower Toarcian records of *Weyla* in eastern Tethyan localities (Turkey, Kenya, Madagascar, etc.) coincide with the beginning of a major transgression. Therefore, the possibility that older marine deposits (potentially containing *Weyla*) are absent or do not include suitable facies cannot be rejected outright. This would weaken the hypothesis of the progressive eastward expansion through the 'Hispanic Corridor' as the only possible route of faunal exchange between eastern Pacific and Tethyan areas. In this connection it is worth noting that the absence of suitable environments for *Weyla* to develop has been inferred to explain its absence in southwest Pacific areas (Damborenea, 1993).

3 While acknowledging that diphyletism seems unlikely, Damborenea and Manceñido (1979, p. 96) recall that the hinge characters of Tethyan species of *Weyla* have neither been illustrated nor seen in museum collections, so that comparison with Pacific species is impossible. On the other hand, Aberhan (1998) argued that European species are different from Pacific ones.

4 The pattern of range expansion described so far implicitly assumes (rather dogmatically) that the faunal spread should be realized along shelf areas. Smith and Tipper (1986) discussed a trans-oceanic spread but discounted this possibility because the width of the Pacific Ocean was considered an effective barrier. However, no information on the prodissoconch of *Weyla* is available, as both Susana Damborenea and Martin Aberhan confirmed by personal communications (respectively September and October 2000): this would allow the type of larval development, planktotrophic or nonplanktotrophic, to be recognized (Jablonski and Lutz, 1983). The biogeographic implications would be of dramatic importance. The sparse information on modern counterparts of Jurassic pectinoids that is available suggests that they possessed planktotrophic

Box 8.1 continued

larvae, though an example of lecithotrophic-type larval shell in a reef-adapted pectinid from the Caribbean has been described (Waller, 1991, 1993).

Other biogeographic arguments on which the hypothesis of the Pliensbachian opening of the 'Hispanic Corridor' has been based rely both on assessments of faunal similarity by using the Simpson coefficient (Hallam, 1983) and on ammonite distribution (Smith and Tipper, 1986; Jakobs, 1995). Similarity has been studied between North and South America and between Europe and each of these areas in an interval from Hettangian–Sinemurian to Aalenian–Bajocian. It is more or less constant between the two Americas, whereas it increases in time between these areas and Europe.

Recently Damborenea (2000) has addressed the problem of the opening and effectiveness of the 'Hispanic Corridor' on the basis of the analysis of faunal similarities of Late Triassic to Middle Jurassic bivalve faunas from cratonic North America, South America, Western Europe and Japan. Similarity has been evaluated using the Simpson coefficient, which stresses similarity (see Chapter 4). Damborenea concluded that the corridor acted as a filter since the Sinemurian, or even the Hettangian. Aberhan (2001) also addressed the same problem by calculating Jaccard and Dice coefficients of faunal similarity of Early Jurassic pectinoid faunas from North American craton, South America, western North American terranes, Europe and North Africa. Aberhan concluded that a 'Hispanic Corridor' was a two-way route which developed from an effective barrier in earlier Early Jurassic times (Hettangian to Sinemurian) into a filter, allowing the passage of five out of 48 selected pectinoid morphotypes during Pliensbachian and Toarcian. The west-to-east tropical transpacific dispersion is rejected, in agreement with Smith and Tipper (1986), because of both the apparent absence of the five morphotypes in western Pacific and eastern Tethys and the large width of the Pacific Ocean. The possible existence of Pacific staging posts is not taken into account. The pre-Bathonian envisaged 'Hispanic Corridor' would have been a shallow seaway at low latitudes and a current system must be assumed to allow the east-to-west transport of larvae or ammonite hatchlings. Aberhan (2001) considered the megamonsoonal circulation for Pangaea a plausible mechanism for seasonally changing

Box 8.1 continued

directions flowing through the corridor. In summary, both Damborenea and Aberhan admit a pre-Bathonian 'Hispanic Corridor' but their conclusions differ with respect to the age of the opening and the number of taxa which dispersed through this hypothetical seaway.

The disjunct distribution of some ammonite genera in the absence of a Pliensbachian 'Hispanic Corridor' has been considered puzzling, as for example the occurrence of *Dubariceras* in both eastern Pacific and Tethys (although with different species), where its easternmost record is in Hungary (Dommergues *et al.*, 1984). Smith and Tipper (1986) explicitly discard the possibility of a trans-pacific dispersion of ammonites enabled by staging posts (a possibility envisaged by both Dommergues *et al.* (1984) and Dommergues (1994)) and implicitly assume ammonites as poorly dispersive organisms, in contrast with the most recent reconstructions of ammonite ecology (Westermann, 1996). Jakobs (1995) has illustrated from localities of the Canadian terranes Wrangellia and Stikinia numerous ammonite species common in Western Tethyan areas. Faunal exchange along Arctic or Austral routes (P_1 and P_2 in Fig. 8.3) is unlikely because at higher latitudes the ecophysiological conditions were unsuitable for these taxa. Jakobs does not reject the hypothesis of the trans-oceanic dispersion embodied in Newton's (1988) model of pantropical faunas but favours the alternative of the 'Hispanic Corridor'. The assumption of adult ammonites dwelling in a shallow sea like the 'Hispanic Corridor' should be rejected because the Tethyan ammonite genera reported from American Cordilleras were not adapted (except *Bouleiceras*) to shallow environments. Aberhan (2001) has noted that the transport of eggs may be a more important factor than the ecological requirements of adults. However, the ecological conditions encountered at the arrival in new areas are also of dramatic importance.

Further biogeographic data are provided by larger benthic foraminifers. As noted in other sections of this book, the larger foraminifers repre-sent a characteristic geoecological assemblage which indicates shallow carbonate platform environments at low latitudes. Post-Triassic larger foraminifers of the family Involutinidae were mostly confined on the southern Tethyan margin during the Lower Jurassic (Bassoullet *et al.*, 1985). In the Pliensbachian their biogeographical range increased and the westermost occurrences are

Box 8.1 continued

recorded in the Moroccan Essaouira Basin and in the Great Bank off Newfoundland. These foraminifers did not reach the Gulf of Mexico before Callovian–Oxfordian times (Fourcade and Michaud, 1987). Their absence could be related to the supposed tectonic unstability of an extensional basin like the 'Hispanic Corridor' in pre-Callovian times. However, the spread of the calcisponge *Stylothalamia*, which is confined to shallow-water carbonate platforms (i.e. the same environment as that of the larger benthic foraminifers) has been postulated, from Western Tethys to Chile and Peru through a Pliensbachian 'Hispanic Corridor' (von Hillebrandt, 1981).

To conclude, the hypothesis of the opening of the 'Hispanic Corridor' during the Pliensbachian is supported by some biogeographic data but geological evidence is still lacking. The faunal similarities between eastern Pacific seas and Western Tethys can be alternatively explained, in agreement with Newton's pantropical model, invoking transpacific dispersion of larvae and, in the case of cephalopods, hatchlings or eggs.

setting with respect to the cratons. Palaeomagnetic and structural studies on terranes have been carried out in order to determine the age of their accretion to the main continental masses and also to reconstruct their original palaeogeographic positions. The geological allochthony of a terrane relative to the autochthonous craton, as deduced from field evidence, and its original palaeogeographic position are distinct, though connected, problems. The latter can be resolved only by taking into account both palaeobiogeographic and palaeomagnetic data.

Palaeontologists are able to recognize that biotic assemblages of terranes clearly display anomalous biogeographic dissimilarities with faunas and floras of the cratons to which they are accreted. Palaeobiogeographical research has produced numerous data that strongly support the theory of the tectonic drift of terranes and their final accretion to main continental masses, although the amplitude of the tectonic transport is controversial. A summary of the numerous case histories cannot be given here, so I will select some examples. However, I refer the reader to both key papers and syntheses on the subject (Monger and Ross, 1971; Tozer, 1982; Ross and Ross, 1985; Hallam,

1986, 1994; Westermann, 1992; Mizutani and Kojima, 1992; Zakharov *et al.*, 1996; Vachard *et al.*, 2000).

8.3.1 Terranes and biogeographical models

Terranes were probably of considerable biogeographic importance because they could have augmented the dispersive potential of organisms. Normally, this depends on the ecological characteristics of the organisms (duration of the larval stage, ecophysiological requirements) and both vectorial and qualitative characteristics of currents. There are two aspects, as follows.

1. Within a Panthalassa as wide as it was during the Palaeozoic and the Early Mesozoic, terranes and islands comparable to modern atolls and guyots may have acted as staging posts within clusters of islands for mollusc and/or coral larvae. Larvae settling in the terranes and reaching the adult stage then generated other larvae that were dispersed by currents to the opposite side of the ocean. This process does not require disruption of environmental equilibrium and could be realized according to the island clustering model. Skelton and Wright (1987) and Skelton (1988) have proposed a similar process to explain the disjunct distribution of the rudist genus *Torreites*, although these authors used the term 'island hopping', which was originally introduced by Simpson (1962) in the context of the colonization of islands by chance dispersal along sweepstakes routes.
2. Tectonic displacement of terranes may cause the transport of biotas according to McKenna's 'Noah's Ark' model. Conversely, distinct successions of biotas have lived in the terrane during the time interval of the terrane drifting and then have become fossils; in this case the terrane may have acted as a 'beached Viking funeral ship' (McKenna, 1973).

8.3.2 Palaeobiogeographic 'anomalies' and terranes

Some classic examples of palaeobiogeographic 'anomalous' distributions associated with allochthonous terranes are summarized below. The methods employed by authors generally consist of the recognition of endemic distributions, use of geoecological assemblages and, in some cases, comparisons by use of similarity coefficients. A critical discussion of these examples will be developed in the next sections.

8.3.2.1 Tethyan fusulinids in North America

One of the most classic examples is that of the distribution of fusulinid genera. Tethyan fusulinids have been recognized in North American terranes that could have drifted from Tethyan areas. The possibility of fusulinid dispersion from eastern Tethyan shelves to North America, across a Panthalassa wider than the modern Pacific Ocean, has been excluded (Stanley and Yancey, 1990). However, the amplitude of the tectonic transport has been questioned, because a Tethyam geographical origin of the terranes in the East Pacific has been envisaged (Tozer, 1982). The occurrence in the cratonal Permian Basin of West Texas of fusulinids considered Tethyan indicators has been interpreted as evidence of wide transpacific dispersion toward the western North American continent (Newton, 1988, 1990). More recent studies based on observations on Recent faunas acknowledge the possibility of dispersion in both east-to-west and west-to-east directions of some fusulinids and other micro-organisms by biological rafts, such as floating algae, also favoured by the presence of stepping stones (Lethiers and Crasquin-Soleau, 1995; Vachard and Fourcade, 1997). On the basis of fusulinid assemblages, Vachard *et al.* (2000) have proposed palaeogeographical positions and time of accretion of numerous Mexican Upper Palaeozoic terranes.

8.3.2.2 Carboniferous and Permian corals

In the terrane of the Cache Creek assemblage (British Columbia, Canada), where Permian fusulinids with Asian affinities have been recognized, Stevens (1983) has described corals of the families Pseudopavonidae (Carboniferous) and Waagenophyllidae (Permian). The westernmost findings of these corals are in Japan and in China for the former family, and in Japan and Indonesia for the latter. Waagenophyllidae are represented in Canada by *Parawentzelella (Myagiella)*, a subgenus known only in Japan. Even in the Asian localities these coral families have been found in terranes. Therefore a slow drift from a 'palaeo-pacific' area towards the eastern and western margins of the modern Pacific Ocean has been envisaged. The accretion of the Cache Creek to the North American craton is dated to Middle Jurassic.

Miller and Wright (1987) reported Late Permian waagenophyllid corals from the Klamath Mountains (California, USA). These corals are associated with Tethyan fusulinids in the Eastern Hayfork terrane but with different, non-Tethyan, fusulinids of North American affinities in the Eastern Klamath terrane. The former terrane is interpreted as a remnant of a subduction complex and the corals occur in exotic blocks; the latter terrane was a part of an island arc and the fossiliferous levels correspond to reefal limestones intercalated in a tuffaceous succession. Miller and Wright reject the hypothesis of tectonic transport from Western Tethys because the required convergence rate of 92 cm/year (calculated on the basis of inferred palaeo-pacific positions and age of accretion) is geologically not realistic. The Hayfork limestone blocks containing typical Tethyan faunas have been interpreted as seamounts that were within the range of Tethyan faunal influences, whereas the Klamath arc has been inferred in a palaeogeographic position where it received faunas from both Tethys and North America. This subduction/island arc system was inferred relatively close to the American craton. If the palaeogeographic position of the terranes has been inferred correctly, the reasons for the occurrence of elements of the Tethyan biota among the North American biota remain to be explained. The terranes probably constituted a network of staging posts that favoured both dispersion and distributional change, as already emphasized for other cases.

8.3.2.3 Triassic bivalves of the Wallowa terrane

The Triassic bivalve fauna from the Wallowa terrane (North-Western USA) is characterized by the presence of variable percentages of taxa showing affinities with different palaeobiogeographic regions (Tethys, North America, South America, Japan/Siberia), together with a significant percentage of endemic species. The most similar pattern, as far as the relative abundances of endemic, Pacific and Indo-Pacific taxa are concerned, is that of an insular community (Newton, 1987) such as modern Hawaiian molluscan communities. The allochthony of the Wallowa terrane is clear but its original palaeogeographic position remains undetermined even if the occurrence of North American–Andean elements allows us to infer relatively short longitudinal transport.

8.3.2.4 Pliensbachian ammonites

In western North America the Pliensbachian ammonite faunas of the Wrangellia, Stikinia and Quesnellia terranes show latitudinal gradients from Tethyan to Boreal biogeographic affinities (Smith and Tipper,

1986). The occurrence of low-latitude bivalves in the same terranes is consistent with the distributional patterns of ammonites (Aberhan, 1998, 1999).

Smith and Tipper used three similarity coefficients (Dice, Jaccard and Simpson) to analyse the relations between the faunas of the terranes and the faunas of both the craton and the Sonomia terrane. The Pliensbachian sediments of the latter can be considered as cratonal because it was accreted to North America in Triassic times. The cratonal areas provide two fixed reference areas: Boreal faunas occur on the craton and Tethyan more diverse faunas in Sonomia. Faunal similarity between these two cratonal areas is low. The similarity coefficients suggest stronger faunal affinities between Sonomia and the terranes than between the craton and the terranes. Furthermore, in the Late Pliensbachian the faunal diversity of the terranes was equal to or greater than the diversity of the Sonomia fauna. However, in the northern parts of the terranes Smith and Tipper distinguished areas where Boreal and Tethyan elements are mixed. Dommergues (1987) considered these faunas with Tethyan affinities as belonging to an East-Pacific Province.

The biogeographic data suggest the reconstruction proposed in Fig. 8.4. Latitudinal displacements of 500 km for Quesnellia, 1800 km for Stikinia and 2400 km for Wrangellia have been suggested. Recent studies on bivalves lower these estimates to 1100–1600 km for Stikinia and 500–1400 km for Wrangellia (Aberhan, 1999). The present-day position of Wrangellia is the result of a large tectonic displacement that brought it north of Stikinia, i.e. latitudinal inversion. The palaeolatitudinal data have been obtained from both palaeomagnetic and palaeobiogeographic evidence. The palaeolongitudes cannot be obtained but the presence of typically East Pacific genera and species, which have never been found in Tethyan and Euro-Boreal areas, leads to the conclusion that these terranes were relatively close to the North American continent.

8.3.2.5 Tithonian ammonites of the Sikhote-Alin terrane

This area is at present located near Vladivostok (far eastern Russia), in the southern part of the Sikhote-Alin range. It was part of the Bureya-Khanka megablock which, according to palaeogeographic reconstructions (Sey *et al.*, 1992), was accreted to the Siberian continent at the end of the Jurassic. In spite of the reconstructed palaeolatitude of about 60°N (Parrish, 1992), the Tithonian ammonite fauna described from Primorje shows common elements with low-latitude Mediterranean faunas (Cecca, 1999). The 'impoverished' character of

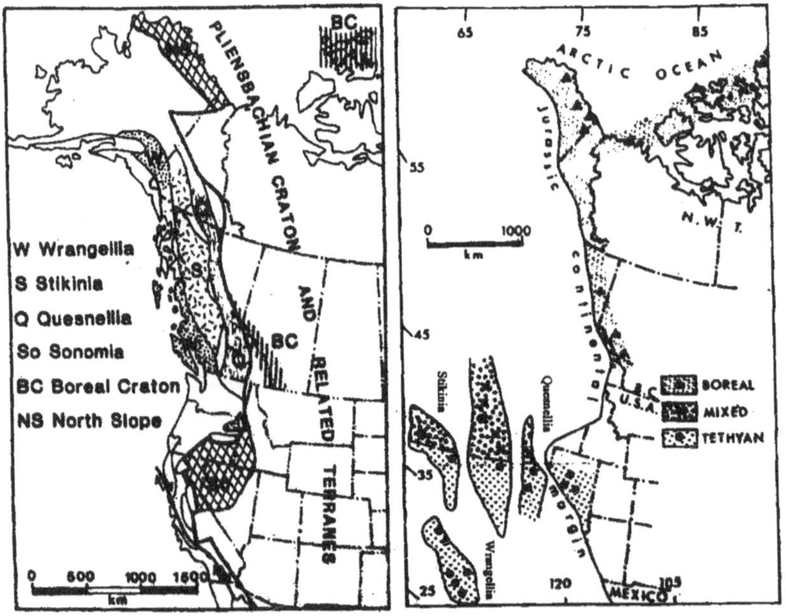

Figure 8.4 Left: structural map of western North America with boundaries of the most important terranes. Right: biogeographic affinities of Pliensbachian ammonites within the Wrangellia, Stikinia and Quesnellia terranes of western North America and the cratonal areas with latitudinal displacement inferred. No relative longitudinal position is implied and possible rotations are not considered.

Source: Smith and Tipper (1986). Modified with permission of the Society for Sedimentary Geology.

the fauna recalls the assemblages of the epicontinental platforms of sub-Mediterranean areas (northern margin of Tethys from the Mediterranean domain to the Middle East), i.e. faunas occurring at palaeolatitudes between 20 and 35°N. The occurrence of buchiids at Primorje, although not abundant, is not *per se* a definite indication of high palaeolatitudes because representatives of this bivalve group occur in other low-latitude localities elsewhere in the world (Hallam, 1994) and display a bipolar distribution (Crame, 1993). The ammonite faunal character is strongly suggestive of a displaced terrane involving northerly tectonic displacement (Hallam, 1986, 1994).

According to the plate reconstruction of Ricou (1994), used in Figs 2.4 and 4.3, a palaeolatitude of about 45°N should apply to Primorje, assuming that the Bureya-Khanka megablock was accreted to the

Siberian continent at the end of the Jurassic (Sey *et al.*, 1992). This palaeolatitude fits better with the palaeobiogeography of the ammonites than the value of about 60°N suggested by Parrish (1992). The low-latitude character of the ammonite fauna of Primorje has been tentatively explained postulating a more southerly position of about 35°N for South Primorje at the end of the Jurassic (I recall that the palaeolongitude cannot be determined). This implies the accretion to the Siberian continent of the Bureya-Khanka block later than previously thought (Cecca, 1999). Accordingly, the occurrence of this Mediterranean assemblage in the Bureya-Khanka terrane would correspond to McKenna's (1973) model of the 'beached Viking funeral ship'.

Zakharov *et al.* (1996) consider the changing latitudinal position of the ecotone (or faunal overlap) of Boreal and Tethyan faunas in the Siberian craton as resulting from climatic control. Accordingly, the Boreal affinities of ammonites and bivalves in the Early Bajocian and in the Early Bathonian followed by the widespread occurrence of Tethyan ammonites (together with the Boreal bivalve *Buchia*) in the Tithonian would be suggestive of a climatic control of the distributional changes rather than a tectonic displacement. The authors conclude that the Sikhote-Alin accreted to the Asian continent in the Early Jurassic. Unfortunately, no independent lines of evidence can be used because no palaeomagnetic data for palaeolatitudes are available.

8.3.3 The amplitude of longitudinal transport of terranes

Geological data demonstrate the allochthony of the terranes recognized on both the eastern and the western sides of the Pacific. Both palaeomagnetic and palaeontological data clearly show, in most cases, evidence of latitudinal displacement. However, the amplitude of the longitudinal transport remains undetermined.

We have discussed the similarity of the Triassic bivalve communities of the Wallowa terrane with the pattern of molluscan faunal composition in Hawaii, in particular the relative proportions of endemics. Let us imagine a fossil equivalent of Hawaiian faunas: the biogeographic relations do not resolve the problem of the correct position of the islands relative to the major continental masses. Because of the absence of North American elements, we could erroneously locate Hawaii close to the western (Asian) Pacific coast, but it actually is, today, exactly in the **middle** of the Pacific Ocean. Seemingly, the Line and Marshall islands, whose faunas are almost devoid of endemics, and are composed of dominant Indo-Pacific taxa plus a minor portion of Pacific taxa, would have been located in the Western, Indo-Pacific part, of the

Pacific area whereas they are also in the middle of the Pacific Ocean. This shows that strong biogeographic affinity cannot be taken on its own as a proof of longitudinal vicinity (or physical connection). The possible dispersion of larvae through passive transport by oceanic currents is complicated by dispersion pathways and by processes of differential settlement in favourable environments which depend on ecophysiological constraints typical of each taxon. On the basis of the occurrence of North American/Andean elements, the Wallowa terrane underwent relatively limited longitudinal transport. According to Newton (1987), it almost certainly occupied a position relatively closer to the North American continent than Hawaii is today. In these islands the lack of North American taxa is probably due to the modern East Pacific barrier, characterized by the absence of staging posts, which limits westward dispersion (Grigg and Hey, 1992).

Taking into account the parallel decreases of surface-water temperature and generic diversity of reef coral faunas in the Pacific, Belasky and Runnegar (1993) argued that temperature gradients and patterns of circulation rather than island stepping stones control the tropical diversity in the eastern Pacific. In fact, the diversity of the Indo-Pacific coral fauna decreases eastwards by a factor of two even in areas with high concentrations of islands; furthermore, the diversity of corals is uniformly high throughout the Indian Ocean, despite the scarcity of islands in its central part. Because IWP hermatypic scleractinians are considered ecological analogues of Palaeozoic Tethyan rugose corals, these authors suggested that the above phenomena were responsible for the existence of longitudinal gradients in coral diversity in the ancient Pacific. Displaced Western North American terranes, such as Cache Creek and Eastern Klamath, contain only a few Tethyan coral taxa, a pattern consistent with an original Eastern Pacific, rather than Western Pacific, location of the terranes.

8.3.4 Methods of reconstruction based on geoecological assemblages

On the basis of analytical studies of geoecological assemblages, different methods have been proposed to evaluate the movements and the amounts of displacement of the terranes relative to cratons. I have selected for discussion the models proposed by Fortey and Cocks (1986) and Aberhan (1998, 1999).

Fortey and Cocks (1986) proposed a model to recognize strike-slip movements and 'mid-oceanic islands', i.e. those elements which can be accreted onto margins of cratons as terranes. Their model deliberately relates palaeolatitude directly to temperature and hence to latitudinally

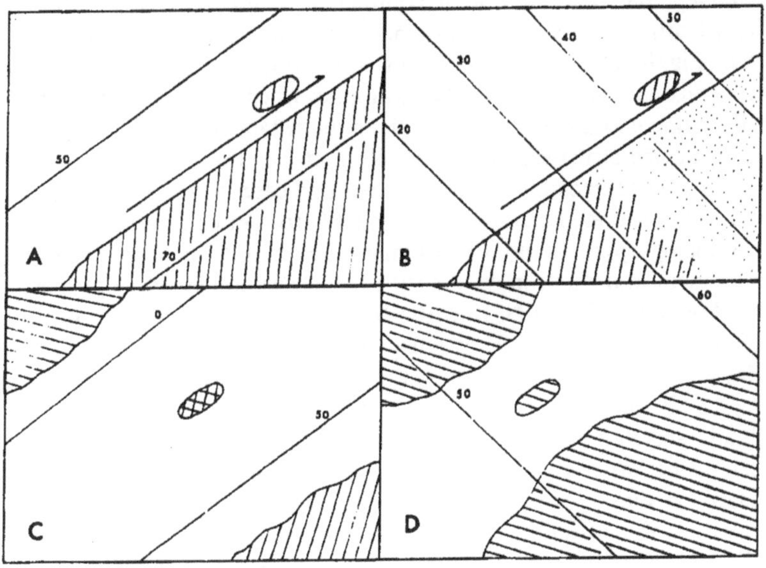

Figure 8.5 Theoretical model to infer the former position of a suspect terrane (the oval island) on the basis of marine faunas. A, strike-slip movement along the same palaeolatitudinal belt; B, across different latitudes; C, mid-ocean island, subsequently accreted, midway between two widely separated continents at different palaeolatitudes; D, between two narrowly separated continents at similar latitudes.

Source: Fortey and Cocks (1986). Reproduced with permission of the authors and the Geological Society.

controlled faunal belts, and does not take into account the disturbance produced by currents. Fig. 8.5 shows four possible palaeogeographic situations. Two of these cases can be detected by using palaeobiogeographic data. The strike-slip movement of an island parallel to palaeolatitude and to a continent edge (Fig. 8.5A) cannot be detected on palaeobiogeographic grounds because the island fauna retains common taxa with the adjacent continent; if the island is displaced obliquely or normal to palaeolatitude (Fig. 8.5B) the movement is detectable if the island moves across-latitude sufficiently to finish adjacent to a different faunal province. The position between two widely separated continents at different palaeolatitudes of a mid-oceanic island subsequently accreted can be recognized on the basis of a 'mixed' fauna resulting from recruitment of distinctive elements of both continental faunas (Fig. 8.5C). However, if the island is located between relatively narrowly separated continents at similar palaeolatitudes, faunal differences are probably inadequate to infer the original position (Fig. 8.5D). Fortey and Cocks (1986) observed that it is easier to recognize such mid-oceanic islands at times when continents are aligned with palaeolatitudinal belts, just the circumstances when it is difficult to recognize strike-slip. It is possible that cases of large-scale movements are undetectable from the faunas and cases where a mid-oceanic island has accreted cannot be recognized from faunas alone.

Biogeographic methods to estimate the amounts of latitudinal displacement and rotation during time have been convincingly proposed by Aberhan (1998, 1999) on the basis of his studies on the history of western Canadian Wrangellia and Stikinia terranes during the Early Jurassic. This is a case resembling the strike-slip movement of Fig. 8.5B, in that terranes move across latitudinal belts to become finally accreted to a craton which was also aligned across latitudinal belts. Aberhan's work confirms the results of Smith and Tipper (1986) based on the analysis of ammonite faunas, although he lowered the values of the latitudinal displacement obtained by Canadian authors. His results are based on different lines of evidence: diversity gradients, faunal similarity, cluster analyses and distributional patterns of characteristic taxa. The biogeographic relations between pectinoid bivalves of these terranes and the cratonic areas of America, as deduced from similarity coefficients, have been discussed in Chapter 4.

Having established the distribution of latitude-dependent species on the craton, the distribution of the same set of species on terranes can be compared. The taxa selected for this kind of study should be biogeographically distinctive, abundant and fairly independent of sedimentary facies. Assuming that distributional limits were more or less parallel to palaeolatitude, the more northerly occurrence of a given species (or a whole latitude-dependent assemblage) identifies on both the craton and the terrane a distributional limit which is roughly parallel to palaeolatitude. If today this limit is strongly offset, the craton and the terrane must have moved relative to each other and the measure of the latitudinal displacement may be approximated from the distance between the two segments of the distributional limit (Fig. 8.6). However, this result takes into account only some elements of the faunas. Minimum and maximum amounts of latitudinal displacement can be suggested when high-palaeolatitude, low-palaeolatitude and mixed faunas are taken into account (Fig. 8.7). The discussed example is based on the case of terranes located in the northern hemisphere, but analogous considerations apply to southern hemisphere terranes.

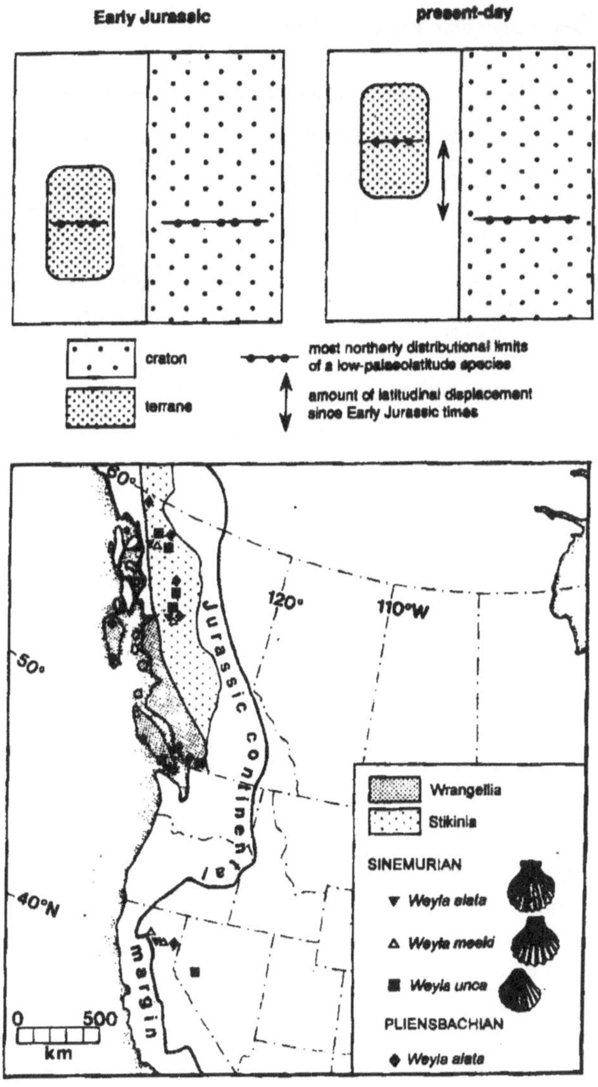

Figure 8.6 Above: theoretical model showing how latitudinal displacement of terranes relative to a craton can be estimated from the distribution of latitude-dependent taxa; the present-day distance between the two segments of the most northerly distributional limit provides an approximate measure of the latitudinal displacement. Below: an actual case study of distribution of three species of the low-latitude bivalve genus *Weyla* during the Liassic on Wrangellia, Stikinia and the North American craton which mirrors the theoretical scheme; note the striking latitudinal differences of most northerly occurrences on the craton compared to those on terranes.

Source: Aberhan (1999). Reproduced by permission of Cambridge University Press.

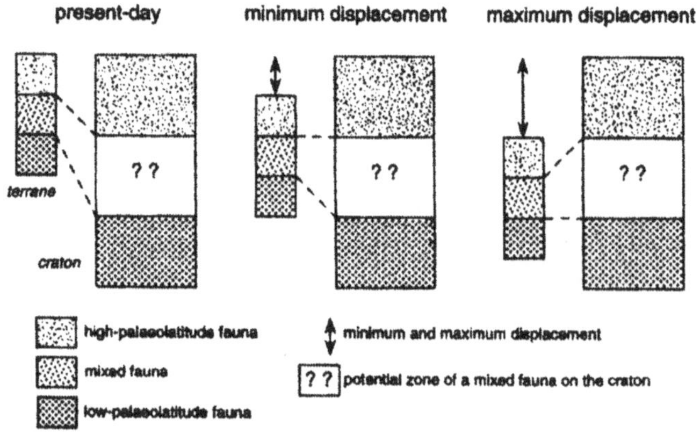

Figure 8.7 Maximum and minimum amounts of latitudinal displacement of a terrane can be estimated approximately on the basis of the boundaries between latitudinally distinct faunas on the terrane and on the craton. Note that on the American craton there is a zone of uncertainty of the precise biogeographic boundaries of Liassic pectinoid faunas.

Source: Aberhan (1999). Reproduced by permission of Cambridge University Press.

8.3.5 The problem of palaeolongitude

Unlike palaeolatitude, which can be determined through both palaeomagnetic and biogeographic data, there is no way of inferring the approximate palaeolongitudinal positions of mobile terranes at different time-horizons. However, both parameters are necessary to fix geographical position. Palaeolongitude may be tentatively inferred from palaeobiogeographic data in certain circumstances.

Belasky and Runnegar (1993) extrapolated the distributions of modern organisms to estimate the longitude, using in particular the eastward decline in living reef coral diversity in the Pacific established by trend-surface analysis (Fig. 8.8). Based on integrating generic diversity of hermatypic corals with palaeomagnetic results, the following 'errors' resulted in placing relative to their actual locations:

1 Easter Island was placed 500 km to the north-west
2 Tahiti was placed 700 km to the north-west
3 Hawaii was placed 3600 km to the south-east.

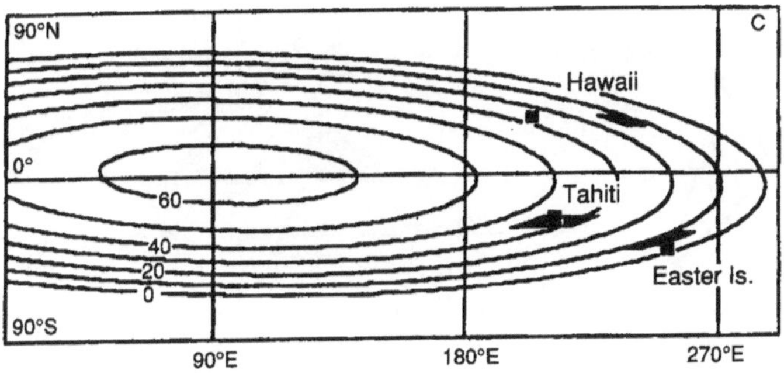

Figure 8.8 Lambert cylindrical equal-area projection of the area that includes the Indian and Pacific Oceans between longitudes 14°E and 66°W. Generic diversity of hermatypic corals is indicated by lines that connect regions with equal numbers of genera. Actual and estimated positions of the islands figured are represented respectively by solid squares and quadrilaterals. Horizontal edges are error bars on palaeomagnetic latitudes; edges subparallel to contours represent error in determination of diversity. The reconstruction is based on a contoured surface of population diversity obtained without diversity data from the eastern Pacific.

Source: Belasky and Runnegar (1993). Reproduced by permission of the Geological Society of America.

The first two can be considered satisfactory estimates, while the large 'error' for Hawaii seems to be due to its anomalously low coral fauna diversity. The longitudinal pattern of decreasing diversity parallels the eastward decrease of surface-water temperature observed in the modern Pacific Ocean. Belasky and Runnegar argued that temperature gradients and patterns of surface-water current circulation, rather than staging posts, control tropical diversity in the eastern Pacific.

Another method was proposed by Aberhan (1998) in order to infer the longitudinal palaeoposition of the Wrangellia and Stikinia terranes in the Early Jurassic. On the basis of the relative affinities, expressed as values of the Dice coefficient of faunal similarity (see Fig. 4.5), between Sinemurian bivalve faunas of these terranes and of both the Western Canada and the Andean cratonic basins, the Andean and Western Canada cratonal areas were further from each other than either was from the terranes (Fig. 8.9, left). This method assumes that the biotic distance can be equated with the geographical distance.

The intersection of equal circles centred in the cratonal basins, whose radius corresponds to their distance, roughly determines the range of probable longitudinal and latitudinal palaeopositions of the terranes.

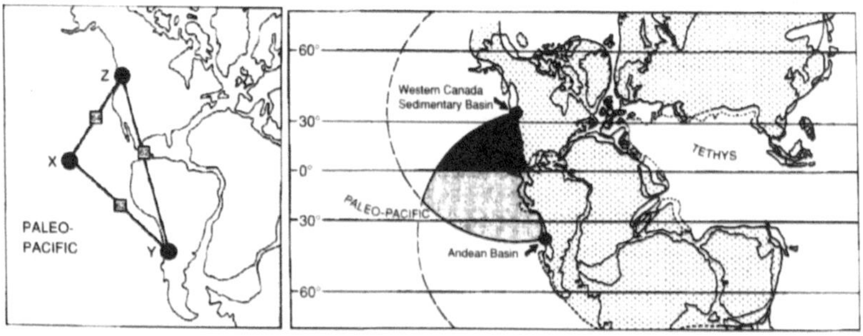

Figure 8.9 Left: the values of the Dice coefficient (in boxes) express the relative affinities between the terranes Wrangellia and Stikinia (X), the Andean (Y) and the Western Canada (Z) cratonal basins; similarities of the terranes to the latter basins are higher than the similarity between the two basins themselves. Right: the probable range of palaeolatitudinal and palaeolongitudinal positions of terranes can be obtained through geometric construction by plotting arcs centred in the cratonal basins, whose radii correspond to the distance YZ; the probable palaeogeographical range corresponds to the grey area if the terranes are in the northern hemisphere, or to the light-grey area if the terranes are in the southern hemisphere.

Source: Aberhan (1998). Modified with permission of the Society for Sedimentary Geology.

Assuming that the terranes were in the northern hemisphere, the range is reduced to the area represented in Fig. 8.9 (right).

8.4 Longitudinal patterns: Newton's model of pantropical faunas

The biogeographic 'anomalies' represented by the occurrence of Tethyan taxa within the faunal assemblages of the North and South American Cordillera have been explained by invoking either their transport over long distance on suspect terranes ('Noah's Ark' model) or 'migration' across marine corridors like the 'Hispanic Corridor' through Pangaea (Fig. 8.10). The historical implications, already discussed, of longitudinal patterns have important applications in palaeogeographic reconstructions (Belasky and Runnegar, 1993).

Newton (1988) proposed an alternative model of steady-state dispersal across the ancient Pacific Ocean, or Panthalassa, to explain fossils of Tethyan affinity occurring in Palaeozoic and lower Mesozoic rocks of the American Cordillera. Tethyan taxa are considered as pantropic species that extended, with attenuated diversities, into Eastern

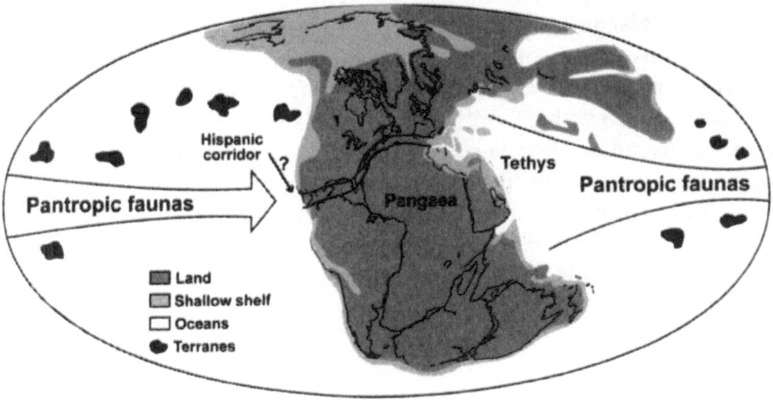

Figure 8.10 The occurrence of Tethyan taxa in North and South American Cordilleras has been explained by the model of the longitudinal displacement of terranes of western Pacific origin. Close to the American continent the proportion of American tropical species would represent an important fraction of the faunal spectrum, whereas a few Tethyan species would occur because of the longitudinal gradient of decreasing Tethyan diversity. The alternative hypothesis is the pre-Bathonian opening of the 'Hispanic corridor'.

Panthalassa. This does not require explanation by either longitudinal tectonic displacement or the formation of marine corridors such as the 'Hispanic Corridor' discussed above.

Tethyan taxa represent only a minor part of the diversity in most Cordilleran faunas, at most 20–25% of the overall species richness of the molluscan associations. Because of both the wide geographical distributions and the small proportion of Tethyan taxa in most Palaeozoic and Mesozoic Cordilleran assemblages, these faunas may have been cosmopolitan tropical taxa that indicate faunal spreads from Tethys or the West Pacific using current systems and island chains as staging posts. The North and South American continents and many of the suspect terranes would have been located on the distal, eastern edge of this tropical belt. However, larval transport by oceanic currents and long durations of the larval stage are required. Alternatively, drifting on biological rafts such as floating algae has been invoked for ostracods and fusulinids. It is worth noting that these different mechanisms permit instantaneous transport relative to geological time.

The dispersal-based pantropic model was not devised as a universal substitute for the longitudinal tectonic displacement or corridor displacement explanations. On the contrary it provides a starting point for palaeobiogeographic testing suggesting a steady-state hypothesis against which other hypotheses can be tested (Newton, 1990). Newton's model was developed to account for 'anomalous' occurrences of Tethyan taxa within Cordilleran assemblages of Palaeozoic to Lower Jurassic terranes. However, evidence of exceptionally wide longitudinal dispersion (it is worth noting that Newton used the term 'dispersal' while envisaging short-term processes of what we call here 'dispersion') has been recognized in several Cretaceous, Cenozoic and modern taxa, in particular corals, rudists and other molluscs. This provides additional evidence for pantropic distributions of pre-Cretaceous Tethyan faunas in the American Cordillera.

8.4.1 Longitudinal decrease of Tethyan influence: a modern example

The Indo West Pacific (IWP) biota provides a modern counterpart for Tethyan biotas. The eastward decrease in the proportion of Tethyan faunas in many Palaeozoic–Mesozoic Cordilleran sites resembles the longitudinal gradient of declining IWP diversity in the islands of the modern Pacific Ocean, which results from dispersion over broad distances. Marine molluscs have a substantial IWP influence: even Easter Island, located far to the east of most Pacific islands, has an attenuated but substantial IWP molluscan fauna. Diversity is attenuated northwards into the Japanese islands, north-eastwards to the Hawaiian islands and eastwards to the Marquesas. IWP influence reaches even into the easternmost Pacific Ocean offshore islands, in particular Clipperton, Galapagos, Revillagigedos, etc. Fifteen IWP molluscan species occur on the western American continental shelf. Hermatypic coral faunas as well as the majority of the non-endemic ahermatypic scleractinians of the Galapagos Islands are predominantly IWP. Newton (1988) has also observed that the occurrence of some IWP taxa off western North America indicates that even in the modern seasonal global climate the influence of this tropical province reaches from the East African margin to western America, some 230° of the globe (approximately 25,000 km). She concluded that there is a fundamental historical link between climatic equability and the geographic range of tropical taxa. Mean modern geographic ranges are probably only 50% to 75% as large as those in earlier, more equable Cenozoic intervals. Extensive pantropic biogeographical distributions for warm-water species during the climatically equable early Mesozoic would be predicted by modern distribution patterns. Conversely, climatic variations can also cause the expansion and contraction of longitudinal distributions of equatorial species. A short-term example is the invasion of the eastern Pacific by IWP species during El Niño events (Grigg and Hey, 1992).

To conclude, the 'anomalous' occurrences of Tethyan fossil species within Cordilleran faunas can be compared with the present-day example of faunas characterized by IWP affinities but living in the eastern Pacific Ocean.

8.4.2 Oceanic surface currents and dispersion: a paradox

The pantropic model implies a wide dispersion of long-lived pelagic larvae before these settle in oceanic islands or atolls that may act as staging posts. Therefore, the configuration and intensity of equatorial currents strongly influences the dispersion of tropical shallow-water organisms between widely separated continental margins or between clusters of islands. Present-day equatorial surface currents in the Pacific Ocean consist of the broad westward Northern and Southern Equatorial Currents together with the intervening narrow Northern Equatorial Countercurrent, which flows eastwards at a latitude of 5° to 10°N. Current velocities range from 25 to 75 cm/s. The main westward flow of these modern equatorial surface currents has been used to infer a westward transpacific spread from Caribbean to mid-oceanic atolls and even to Eastern Tethys, particularly during the Cretaceous. Mid-Cretaceous Texas–Mexican fossil reef corals have been reported from the summits of western Pacific guyots, which were located in the eastern Central Pacific near the Equator at that time. Turonian to Maastrichtian shallow water foraminifera with Caribbean affinities have been reported from the Line and Marshall islands (Schlanger *et al.*, 1981). Skelton (1988) inferred a westerly transpacific dispersion of Campanian rudists from the Caribbean to Oman, favoured by staging posts (but a different hypothesis was put forward by Philip and Platel (1994) and Philip (1998). A westward surface current has been inferred to support the idea of likely dispersion from Western Tethys to Pacific via the 'Hispanic corridor'. Although the eastward equatorial countercurrent is theoretically plausible, dispersion from Tethys to eastern Pacific has not been taken into account by some authors (Stevens, 1983; Ross and Ross, 1985; Smith and Westermann, 1990; Stanley and Yancey, 1990). However, data on present-day Pacific molluscan assemblages unequivocally demonstrate that IWP taxa dispersed eastwards to the eastern Pacific. No westward dispersion of American mollusc species to the IWP areas has been observed. This is a paradox because currents are clearly the physical vector controlling larval dispersion (recruitment and eventual settlement depending on other, distinct, factors).

It has been suggested that easterly-flowing equatorial surface countercurrents are responsible for this eastward transport of IWP faunas. For example, the Cromwell current in the Pacific Ocean is more rapid than surface currents and flows eastwards with velocities that range from 100 to 150 cm/s along the Equator. The average width of this jetlike equatorial undercurrent is only 300 km; it occurs at depths from 50 to 300 m with maximum velocities at 75 to 100 m and its core generally coincides with the thermocline. Therefore, eastern transport of IWP faunas across the modern Pacific may be linked to the eastward Cromwell undercurrent, thus removing the paradox of the westerly flow of the equatorial surface currents which contrasts with the observed biogeographic distributional patterns. Even in the Atlantic Ocean, the equatorial undercurrent contains abundant larvae of Cnidaria, Echinodermata, Arthropoda, Annelida and Mollusca (Scheltema, 1968).

Concluding remarks

A book conceived as a textbook generally has no place for conclusions, at least not of the kind we usually read in, or write for, scientific articles. Thus, the following remarks are aimed at outlining possible perspectives for further research in palaeobiogeography.

The debate which classically opposes vicariance and dispersal biogeography is a continuing one. However, a sort of happy medium or compromise has been achieved by some vicariance biogeographers who do not reject dispersal and are prepared to admit that dispersal and vicariance may occur at different times (see the discussion of Liebermann's geodispersal in Chapter 2).

An important and crucial problem, as sympatric models are increasingly invoked, is the geographic dimension of speciation. Norris (2000) stresses the importance of sympatric speciation in pelagic organisms such as foraminifers driven by a change in the place of reproduction in, for example, different parts of the water column (models reflecting depth parapatry and depth sympatry). However, using the same data others have arrived at different views. The most striking example is that of Schneider and Kennett (1999), who consider that depth segregation in planktonic foraminifers follows an allopatric rather than a sympatric model (see Chapter 3). A further difficulty is that inferred sympatric speciation is sometimes not marked by any significant morphological changes (De Vargas *et al.*, 2001). This can be seen as justifying a degree of taxonomic oversplitting (Norris, 2000). The debate opposing sympatric and allopatric models has important implications for phylogenesis and is relevant to cladistic biogeography. Palaeobiogeographical patterns alone cannot resolve this kind of problem. Despite the inferred sympatric speciation of some groups, areas of endemism really exist. A possible methodology to address such questions is to analyse evolutionary radiation in terms of the biogeographical distribution of the different morphologies, information that should be in place before examining the biogeographical distribution of the taxa (the phyletic units). This approach was adopted by Dommergues *et al.* (1996, 2001). Notwithstanding the taxonomic status of the different morphological types, these authors initially studied the morphological disparity of Liassic ammonites in relation to the major Early Jurassic palaeogeographical events (Dommergues *et al.*, 1996). They then defined the palaeobiogeographical distribution of these morphotypes rather than the distribution of the different taxa (Dommergues *et al.*, 2001). The next step is the phylogenetic reconstruction of the different taxa and the delineation of their biogeographic distributions.

The reluctance of palaeontologists to generate cladograms for fossil species has considerably limited the impact of cladistic, or cladistic-derived, techniques in palaeobiogeography. Two analytical methods, BPA and PAE, can be applied to problems of palaeogeographic reconstruction. The former needs a previously established phylogeny of the fossil groups concerned, whereas PAE can be applied to groups whose phylogenetic relationships are unspecified. It would be interesting to compare the results of these two methods using the same data, but neither has been extensively exploited so far.

There is a wide array of possible interactions between palaeobiogeography and palaeogeography. Palaeobiogeography has been applied to palaeogeographic and palaeotectonic reconstructions (examples are discussed in Chapter 8) but can be useful in other fields reconstructing Earth history. One of the most important aspects of biogeography is the search for a general theory of diversity in space and time. The most important feature of global scale biogeography in the Modern Earth is the latitudinal biodiversity gradient, peaking in the Tropics and decreasing towards the Poles. This is believed to be a perennial characteristic of the biosphere. However, it is important to know whether or not this gradient has always been as steep as it is today. From a biogeographical perspective, temporal fluctuations in biodiversity should be assessed through the establishment of gradients for different periods. This dynamic perspective has not yet been seriously exploited by palaeobiogeographers. The fossil record and the discontinuous nature of exposures of the different geological horizons do not allow the use of quantitative techniques of investigation comparable with those used in neobiogeography. These clearly aid investigations of the relative roles of factors that disrupt the latitudinal pattern of biodiversity. For example, O'Brien (1993, 1998) was able to demonstrate the role of climate and climate change in controlling patterns of diversity of woody plants. Among marine organisms, the role of climate in controlling biogeographical latitudinal and longitudinal patterns of diversity is illustrated by the biogeographic distribution of characteristic geoecological assemblages of warm and shallow-water environments, such as those of carbonate platforms. Crame (1993) also envisaged vicariant events driven by climatic changes, but in a number of fossil groups the climatic perspective has yet to be exploited.

Apart from quantitative techniques of defining bioprovinces (so-called 'phenetics'), the incompleteness of the fossil record has certainly hampered the quantitative approach to palaeobiogeography. In this context, I particularly enjoy a remark of Rosenzweig (1995, pp. 56–57):

> I think paleobiologists have been too cautious and slow in applying the census techniques of ecologists. They are held back by the assumptions of the techniques, and the realization that those assumptions are either difficult to prove for fossil sequences, or else wrong. I have a secret to tell. Those of us who census live animals face exactly the same difficulties. But we work onward. After all, no science is perfect, but scientists proceed in spite of that, aware that they haven't done too badly with the inexact tools that have been available.

I am one of those palaeobiologists who have been too cautious and slow in applying quantitative techniques.

Rosenzweig's comment is both a reassuring admission and an encouragement. I hope that all palaeontologists interested in the kind of problems dealt with in this book will be stimulated and, with the inexact tools available, will carry our science forward.

Bibliography

Aberhan M. (1994) Guild-structure and evolution of Mesozoic benthic shelf communities. *Palaios*, **9**, 516–545.

Aberhan M. (1998) Paleobiogeographic patterns of pectinoid bivalves and the Early Jurassic tectonic evolution of western Canadian terranes. *Palaios*, **13**, 129–148.

Aberhan M. (1999) Terrane history of the Canadian Cordillera: estimating amounts of latitudinal displacement and rotation of Wrangellia and Stikinia. *Geol. Mag.*, **136**, 481–492.

Aberhan M. (2001) Bivalve palaeobiogeography and the Hispanic Corridor: time of opening and effectiveness of a proto-Atlantic seaway. *Palaeogeogr. Palaeoclimat. Palaeoecol.*, **165**, 375–394.

Aberhan M. and Fürsich F. T. (1997) Diversity analysis of Lower Jurassic bivalves of the Andean Basin and the Pliensbackian–Toarcian mass extinction. *Lethaia*, **29**, 181–195.

Aberhan M. and Fürsich F. T. (2000) Mass origination versus mass extinction: the biological contribution to the Pliensbachian–Toarcian extinction event. *Journ. Geol. Soc., London*, **157**, 55–60.

Adams C. C. (1902) Southeastern United States as a center of geographical distribution of flora and fauna. *Biol. Bull.*, **3**, 115–131.

Améziane-Cominardi N., Bourseau J.-P. and Roux M. (1987) Les crinoïdes pédonculés de Nouvelle-Calédonie (S. W. Pacifique): une faune bathyale ancestrale issue de la Mésogée mésozoïque. *C. R. Acad. Sciences Paris*, sér. III, **304**, 15–18.

Archer A. W. and Maples C. G. (1987) Monte Carlo simulation of selected binomial similarity coefficients (I): effect of number of variables. *Palaios*, **2**, 609–617.

Arkell W. J. (1956) *Jurassic Geology of the World*. Oliver & Boyd, Edinburgh.

Armstrong D. M. (1977) Dispersal vs. dispersion: process vs. pattern. *Syst. Zool.*, **26**, 210–211.

Babinot J.-F. and Colin J.-P. (1992) Marine ostracode provincialism in the Late Cretaceous of the Tethyan realm and the Austral Province. *Palaeogeogr. Palaeoclimat. Palaeoecol.*, **92**, 283–293.

Ball I. R. (1976) Nature and formulation of biogeographical hypotheses. *Syst. Zool.*, **24**, 407–430.

Bambach R. K. (1977) Species richness in marine benthic habitats through the Phanerozoic. *Paleobiology*, **3**, 152–167.

Bambach R. K. (1985) Classes and adaptive variety: the ecology of diversification in marine faunas through the Phanerozoic. In: Valentine J. W. (Ed.), *Phanerozoic Diversity Patterns*, 191–253. Princeton University Press, Princeton, NJ.

Bambach R. K. (1993) Seafood through time: changes in biomass, energetics, and productivity in the marine ecosystem. *Paleobiology*, **19** (3), 372–397.

Bambach R. K., Scotese C. R. and Ziegler M. A. (1980) Before Pangaea: the geographies of the Palaeozoic world. *Am. Sci.*, **68**, 26–38.

Barron E. J. and Peterson W. H. (1989) Model simulation of the Cretaceous ocean circulation. *Science*, **244**, 684–686.

Bartolini A. and Larson R. (2001) Pacific microplate and the Pangea supercontinent in the Early–Middle Jurassic. *Geology*, **29**, 735–738.

Bassoullet J.-P., Elmi S., Poisson A., Ricou L.-E., Cecca F., Bellion Y., Guiraud R. and Baudin F. (1993) Middle Toarcian (184 to 182 Ma). In: Dercourt J., Ricou L. E. and Vrielynck B. (Eds), *Atlas Tethys Palaeoenvironmental Maps*. Explanatory Notes. Gauthier-Villars, Paris, 63–80; map 1/20.000.000 by Bassoullet J.-P., Poisson A., Elmi S., Cecca F., Bellion Y., Guiraud R., Le Nindre J. Y. and Manivit J. – Middle Toarcian Paleoenvironments. BEICIP-FRANLAB, Rueil-Malmaison.

Bassoullet J.-P., Fourcade E. and Peybernès B. (1985) Paléobiogéographie des grands Foraminifères benthiques des marges néo-téthysiennes au Jurassique et au Crétacé inférieur. *Bull. Soc. Géol. Fr.*, sér. 8, **I** (5), 699–713.

Baud A., Marcoux J., Giraud R., Ricou L.-E. and Gaetani M. (1993) Late Murghabian Palaeoenvironments (266 to 264 Ma). In: Dercourt J., Ricou L. E. and Vrielynck B. (Eds), *Atlas Tethys Palaeoenvironmental Maps*. BEICIP-FRANLAB, Rueil-Malmaison.

Belasky P. (1992) Assessment of sampling bias in biogeography by means of a probabilistic estimate of taxonomic diversity: application to modern Indo-Pacific reef corals. *Palaeogeogr. Palaeoclimatol. Palaeoecol.*, **99**, 243–270.

Belasky P. (1996) Biogeography of Indo-Pacific larger foraminifera and scleractinian corals: a probabilistic approach to estimating taxonomic diversity, faunal similarity, and sampling bias. *Palaeogeogr. Palaeoclimatol. Palaeoecol.*, **122**, 119–141.

Belasky P. and Runnegar B. (1993) Biogeographic constraints for tectonic reconstructions of the Pacific region. *Geology*, **21**, 979–982.

Bennett K. D. (1990) Milankovitch cycles and their effects on species in ecological and evolutionary time. *Paleobiology*, **16**, 11–21.

Benton M. J., Wills M. A. and Hitchin R. (2000) Quality of the fossil record through time. *Nature*, **403**, 534–537.

Bernasconi M. P. and Robba E. (1982) The Thecosomatous Pteropods: a contribution toward the Cenozoic Tethyan paleobiogeography. *Boll. Soc. Pal. It.*, **21**, 211–222.

Bhaud M. (1989) Rôle de la dissémination larvaire en paléobiogéographie réévalué à la lumière des données concernant l'époque actuelle. *Bull. Soc. Géol. France*, sér. 8, **5** (3), 551–559.

Blondel J. (1979) *Biogéographie et Écologie*. Collection d'Ecologie, 15, Masson, Paris.

Bouchet P. and Taviani M. (1992) The Mediterranean deep-sea fauna: pseudopopulations of Atlantic species? *Deep-Sea Research*, **39**, 169–184.

Boucot A. J. (1975a) Standing diversity of fossil groups in successive intervals of geologic time viewed in the light of changing levels of provincialism. *J. Paleont.*, **49**, 1105–1111.

Boucot A. J. (1975b) *Evolution and Extinction Rate Controls*. Elsevier, Amsterdam.

Boucot A. J. and Gray J. (1983) A Paleozoic Pangaea. *Science*, **222** (4624), 571–581.

Bremer K. (1992) Ancestral areas: a cladistic reinterpretation of the center of origin concept. *Syst. Biol.*, **41**, 436–445.

Bretsky P. W. (1973) Evolutionary patterns in the Paleozoic Bivalvia: documentation and some theoretical considerations. *Geol. Soc. Amer. Bull.*, **84**, 2079–2096.

Briggs J. C. (1974) *Marine Zoogeography*. McGraw-Hill, New York.

Briggs J. C. (1981) Do centers of origin have a center? *Paleobiology*, **7**, 305–307.

Briggs J. C. (1995) *Global Biogeography*. Developments in Palaeontology and Stratigraphy, **14**, Elsevier, Amsterdam.

Brooks D. R. (1981) Hennig's parasitological method: a proposed solution. *Syst. Zool.*, **30** (3), 229–249.

Brooks D. R. (1985) Historical ecology: a new approach to studying the evolution of ecological associations. *Ann. Missouri Bot. Gard.*, **72**, 660–680.

Brooks D. R. (1990) Parsimony analysis in historical biogeography and coevolution: methodological and theoretical update. *Syst. Zool.*, **39** (1), 14–30.

Brown J. H. (1988) Species diversity. In: Myers A. A. and Giller P. S. (Eds), *Analytical Biogeography: an Integrated Approach to the Study of Animal and Plant Distributions*, 57–89. Chapman & Hall, London.

Brundin L. (1988) Phylogenetic biogeography. In: Myers A. A. and Giller P. S. (Eds), *Analytical Biogeography: an Integrated Approach to the Study of Animal and Plant Distributions*, 343–369. Chapman & Hall, London.

Budd A. F., Stemann T. A. and Johnson K. G. (1994) Stratigraphic distributions of genera and species of Neogene to Recent Caribbean reef corals. *J. Paleont.*, **68**, 951–977.

Budd A. F., Stemann T. A. and Stewart R. H. (1992) Eocene Caribbean reef corals: a unique fauna from the Gatuncillo Formation of Panama. *J. Paleont.*, **66**, 570–594.

Callomon J. H. (1985) The evolution of the Jurassic ammonite family Cardioceratidae. *Spec. Papers in Palaeontology*, **33**, 49–90.

Campbell C. A. and Valentine J. W. (1977) Comparability of modern and ancient marine faunal provinces. *Paleobiology*, **3**, 49–57.

Cavalli-Sforza L. L., Menozzi P. and Piazza A. (1997) *Storia e Geografia dei Geni Umani*. Biblioteca Scientifica Adelphi, **25**, Adelphi Edizioni S. p. A., Milan.

Cecca F. (1986) *Le Tithonique de la bordure ardéchoise dans la région du stratotype de l'Ardescien: étude stratigraphique et paléontologique*. Thèse Dipl. Doct. Univ. Claude-Bernard, Lyon I, n. 20–86, 1–263.

Cecca F. (1988) Ammonites méditerranéennes du Tithonique inférieur de l'Ardèche (Sud-Est de la France): analyse des afflux téthysiens. *Geobios*, **21**, 169–186, Lyon.

Cecca F. (1992) Ammonite habitats in the Early Tithonian of Western Tethys. *Lethaia*, **25**, 257–267.

Cecca F. (1998) Early Cretaceous (pre-Aptian) ammonites of the Mediterranean Tethys: Palaeoecology and Palaeobiogeography. *Palaeogeogr. Palaeoclimat. Palaeoecol.*, **138**, 305–323.

Cecca F. (1999) Palaeobiogeography of Tethyan ammonites during the Tithonian (latest Jurassic). *Palaeogeogr. Palaeoclimat. Palaeoecol.*, **147**, 1–37.

Cecca F. and Enay R. (1991) Les Ammonites des zones à semiforme et à fallauxi du Tithonique de l'Ardèche (Sud-Est de la France): stratigraphie, paléontologie, paléobiogéographie. *Palaeontographica Abt. A*, **219**, 1–87.

Censky E. J., Hodge K. and Dudley J. (1998) Over-water dispersal of lizards due to hurricanes. *Nature*, **395**, 556.

Cheetham A. H. and Hazel J. E. (1969) Binary (presence–absence) similarity coefficients. *J. Paleontol.*, **43**, 1130–1136.

Chirat R. (2000) The so-called 'cosmopolitan palaeobiogeographic distribution' of Tertiary Nautilida of the genus *Aturia* Bronn 1838: the result of post-mortem transport by oceanic palaeocurrents. *Palaeogeogr. Palaeoclimat. Palaeoecol.*, **157**, 59–77.

Christiansen J. L. and Stouge S. (1999) Oceanic circulation as an element in palaeogeographical reconstructions: the Arenig (early Ordovician) as an example. *Terra Nova*, **11**, 73–78.

Coates A. G. (1973) Cretaceous Tethyan coral–rudist biogeography related to the evolution of the Atlantic Ocean. *Spec. Papers in Palaeontology*, **12**, 169–174.

Cocks L. R. M. and Fortey R. A. (1982) Faunal evidence for oceanic separations in the Palaeozoic of Britain. *Jour. Geol. Soc., London*, **139**, 465–478.

Collina-Girard J. (1999) Observation en plongée de replats d'érosion eustatique à l'île d'Elbe (Italie) et à Marie-Galante (Antilles): une séquence bathymétrique mondiale? *C. R. Acad. Sci. Paris*, sér. II, **328**, 823–829.

Congrès sur les méthodes de la stratigraphie (1972) Discussion à la suite des rapports de Monsieur Tintant et de Madame Le Calvez. Mém. B. R. G. M., **77**, 41–44.

Cooper R. A., Fortey R. A. and Lindholm K. (1991) Latitudinal and depth zonation of early Ordivician graptolites. *Lethaia*, **24**, 199–218.

Cracraft J. (1980) Biogeographic patterns of terrestrial vertebrates in the Southwest Pacific. *Palaeogeogr. Palaeoclimat. Palaeoecol.*, **31**, 353–369.

Crame J. A. (1992) Evolution history of the polar regions. *Historical Biology*, **6**, 37–60.

Crame J. A. (1993) Bipolar molluscs and their evolutionary implications. *Journal of Biogeography*, **20**, 145–161.

Crame J. A. (1996) Antarctica and the evolution of taxonomic diversity gradients in the marine realm. *Terra Antarctica*, **3**, 121–134.

Crame J. A. (2000) Evolution of taxonomic diversity gradients in the marine realm: evidence from the composition of Recent bivalve faunas. *Paleobiology*, **26**, 188–214.

Craw R. (1988) Panbiogeography: method and synthesis in biogeography. In: Myers A. A. and Giller P. S. (Eds), *Analytical Biogeography: an Integrated Approach to the Study of Animal and Plant Distributions*, 405–435. Chapman & Hall, London.

Crisp D. J. (1976) Energy relations of marine invertebrate larvae. *Thalassia Jugoslavica*, **10**, 102–120.

Croizat L. (1952) *Manual of Phytogeography*. W. Junk, The Hague.

Croizat L. (1958) *Panbiogeography*, 3 vols. Published by the author, Caracas.

Croizat L. (1962–1964) *Space, Time, Form: the Biological Synthesis*. Published by the author, Caracas.

Croizat L. (1982) Vicariance/vicariism, panbiogeography, 'vicariance biogeography', etc.: a clarification. *Syst. Zool.*, **31**, 291–304.

Crovello T. J. (1981) Quantitative biogeography: an overview. *Taxon*, **30**, 563–575.

Damborenea S. E. (1993) Early Jurassic South American pectinaceans and circum-Pacific palaeobiogeography. *Palaeogeogr. Palaeoclimatol. Palaeoecol.*, **100**, 109–123.

Damborenea S. E. (1998) The bipolar bivalve *Kolymonectes* in South America and the diversity of Propeamussiidae in Mesozoic times. In: Johnston P. A. and Haggart J. W. (Eds), *An Eon of Evolution – Paleobiological Studies Honoring Norman D. Newell*, 143–155, University of Calgary Press.

Damborenea S. E. (2000) Hispanic Corridor: its evolution and the biogeography of bivalve molluscs. *GeoResearch Forum*, **6**, 369–380.

Damborenea S. E. and Manceñido M. O. (1979) On the palaeogeographical distribution of the pectinid genus *Weyla* (Bivalvia, Lower Jurassic). *Palaeogeogr. Palaeoclimatol. Palaeoecol.*, **27**, 85–102.

Darlington P. J. (1959) Darwin and zoogeography. *Proc. Amer. Phil. Soc.*, **103**, 307–319.

de Candolle A. P. (1820) Géographie botanique. *Dict. des Sciences Natur.*, **18**, 359–422.

De Vargas C., Renard S., Hilbrecht H. and Pawlowski J. (2001) Pleistocene adaptive radiation in *Globorotalia truncatulinoides*: genetic, morphologic, and environmental evidence. *Paleobiology*, **27**, 104–125.

Dercourt J., Ricou L. E. and Vrielynck B. (Eds) (1993) *Atlas Tethys Palaeoenvironmental Maps*. Gauthier-Villars, Paris.

Dhondt A. V. (1992) Cretaceous inoceramid biogeography: a review. *Palaeogeogr. Palaeoclimatol. Palaeoecol.*, **92**, 217–232.

Dieckmann U. and Doebeli M. (1999) On the origin of species by sympatric speciation. *Nature*, **400**, 354–357.

Dommergues J.-L. (1987) L'évolution chez les Ammonitina du Lias moyen (Carixien, Domérien basal) en Europe occidentale. *Docum. Lab. Géol. Lyon*, **5** (4), 355–407.

Dommergues J.-L. (1994) Les faunes d'ammonites pandémiques et téthysiennes du Lias confrontées à un modèle de distribution pantropicale. *Palaeopelagos, Spec. Publ.*, **1**, 93–107.

Dommergues J.-L., Laurin B. and Meister C. (1996) Evolution of ammonoid morphospace during the Early Jurassic radiation. *Paleobiology*, **22**, 219–240.

Dommergues J.-L., Laurin B. and Meister C. (2001) The recovery and radiation of Early Jurassic ammonoids: morphologic versus palaeobiogeographical patterns. *Palaeogeogr. Palaeoclimat. Palaeoecol.*, **165**, 195–213.

Dommergues J.-L. and Marchand D. (1988) Paléobiogéographie historique et écologique: application aux Ammonites du Jurassique. In Wiedmann J. and Kullmann J. (Eds), *Cephalopods – Present and Past*, 351–364. Schweizerbart' sche Verlagsbuchhandlung, Stuttgart.

Dommergues J.-L., Marchand D., Sapunov I. and Thierry J. (1987a) Les faunes d'ammonites du Jurassique bulgare: une confirmation paléobiogéographique de l'isolement relatif de la région balkano-moesienne. *Bull. Soc. Géol. France* sér. 8, **3** (4), 737–742.

Dommergues J.-L., Marchand D. and Thierry J. (1987b) Biogéographie des ammonites jurassiques et reconstitution palinspastique de la Téthys. *Geodinamica Acta*, **1**, 273–281.

Dommergues J.-L., Mouterde R. and Rivas P. (1984) Un faux polymorphitiné: *Dubariceras*, nouveau genre d'ammonites du Carixien mésogéen. *Geobios*, **17**, 831–839.

Doyle P. (1987) Lower Jurassic–Lower Cretaceous belemnite biogeography and the development of the Mesozoic Boreal Realm. *Palaeogeogr. Palaeoclimatol. Palaeoecol.*, **61**, 237–254.

Doyle P. (1992) A review of the biogeography of Cretaceous belemnites. *Palaeogeogr. Palaeoclimatol. Palaeoecol.*, **92**, 207–216.

Ekman S. (1953) *Zoogeography of the Sea*. Sidgwick & Jackson, London.

Eldredge N. (1981) Discussion of M. D. F. Udvardy's paper. In Nelson G. and Rosen D. E. (Eds), *Vicariance Biogeography: a Critique*, 34–38. Columbia University Press, New York.

Eldredge N. (1995) *Reinventing Darwin. The Great Debate at the High Table of Evolutionary Theory*. John Wiley & Sons, New York.

Elmi S. (1993) Les voies d'échange faunique entre l'Amérique du Sud et la Téthys alpine pendant le Jurassique inférieur et moyen. *Docum. Lab. Géol. Lyon*, **125**, 139–149.

Enay R. (1972) Paléobiogéographie des Ammonites du Jurassique terminal (Tithonique/Volgien/Portlandien l.s.) et mobilité continentale. *Geobios*, **5**, 355–407.

Enay R. (1980a) Indices d'émersion et d'influences continentales dans l'Oxfordien supérieur – Kimméridgien inférieur en France. Interprétation paléogéographique et conséquences paléobiogéographiques. *Bull. Soc. Géol. France*, sér. 7, **22** (4), 581–590.

Enay R. (1980b) Paléobiogéographie et Ammonites jurassiques: 'rythmes fauniques' et variations du niveau marin; voies d'échanges, migrations et domaines biogéographiques. Livre jubilaire de la Soc. Géol. France, 1830–1980. *Mém. H. Sér. Soc. Géol. France*, **10**, 261–281.

Enay R. (1985) Seuils et voies d'échange: rôle dans la distribution et l'évolution des ammonites jurassiques. *Min. Educ. Nat. Com. Trav. Hist. Scient., Bull. Sect. Sci.*, **9**, 203–215.

Enay R., Bellion Y., Cariou E., Guiraud R., Mangold C., Ricou L. E. and Thierry J. (1993) Callovian. In: Dercourt J., Ricou L. E. and Vrielynck B. (Eds), *Atlas Tethys Palaeoenvironmental Maps*. Maps. BEICIP-FRANLAB, Rueil-Malmaison.

Enay R. and Cariou E. (1997) Ammonites faunas and palaeobiogeography of the Himalayan belt during the Jurassic: initiation of a Late Jurassic austral ammonite fauna. *Palaeogeogr. Palaeoclimat. Palaeoecol.*, **134**, 1–38.

Enay, R. and Mangold C. (1982) Dynamique biogéographique et évolution des faunes d'Ammonites au Jurassique. *Bull. Soc. Géol. France*, sér. 7, **24** (5–6), 1025–1046.

Erdmann M. V., Caldwell R. L. and Moosa M. K. (1998) Indonesian 'king of the sea' discovered. *Nature*, **395**, 335.

Etter W. (1999) Community analysis. In: Harper D. A. T. (Ed.), *Numerical Palaeobiology*, 285–360. John Wiley and Sons, New York.

Fallaw W. C. (1979) A test of the Simpson coefficient and other binary coefficients of faunal similarity. *J. Paleont.*, **53**, 1029–1034.

Fallaw W. C. and Dromgoole E. L. (1980) Faunal similarities across the South Atlantic among Mesozoic and Cenozoic invertebrates correlated with widening of the ocean basin. *Journ. Geology*, **88**, 723–727.

Fernandez-Lopez S. (1991) Taphonomic concepts for a theoretical biochronology. *Rev. Esp. Paleont.*, **6**, 37–49.

Fitch W. M. (1971) Toward defining the course of evolution: minimum change for a specific tree topology. *Syst. Zool.*, **200**, 406–416.

Flessa K. W. (1975) Area, continental drift and mammalian diversity. *Paleobiology*, **1**, 189–194.

Flessa K. W. and Jablonski D. (1996) The geography of evolutionary turnover: a global analysis of extant bivalves. In Jablonski D., Erwin D. H. and Lipps J. H. (Eds), *Evolutionary Paleobiology*, 376–397. The University of Chicago Press, Chicago and London.

Flessa K. W. and Thomas R. H. (1985) Modeling the biogeographic regulation of evolutionary rates. In: Valentine J. W. (Ed.), *Phanerozoic Diversity Patterns*, 355–376. Princeton University Press, Princeton, NJ.

Foote M. and Sepkoski J. J. Jr (1999) Absolute measures of the completeness of the fossil record. *Nature*, **398**, 415–417.

Forey P. (1998) A home from home for coelacanths. *Nature*, **395**, 319–320.

Fortey R. A. and Cocks L. R. M. (1986) Marginal faunal belts and their structural implications, with examples from the Lower Palaeozoic. *Jour. Geol. Soc., London*, **143**, 151–160.

Fortey R. A. and Mellish C. J. T. (1992) Are some fossils better than others for inferring palaeogeography? The early Ordovician of the North Atlantic region as an example. *Terra Nova*, **4**, 210–216.

Fortey R. A., Owens R. M. and Rushton A. W. A. (1989) The palaeogeographic position of the Lake District in the Early Ordovician. *Geological Magazine*, **126**, 9–17.

Fourcade E., Azéma J., Cecca F., Bonneau M., Peybernès B. and Dercourt J. (1991) Essai de reconstitution cartographique de la paléogéographie et des paléoenvironnements de la Téthys au Tithonique supérieur (138 à 135 Ma). *Bull. Soc. Géol. France*, **162** (6), 1197–1208.

Fourcade E., Azéma J., Cecca F., Dercourt J., Guiraud R., Sandulescu M., Ricou L. E., Vrielynck B., Cottereau N. and Petzold M. (1993) Late Tithonian (138 to 135 Ma). In: Dercourt J., Ricou L. E. and Vrielynck B. (Eds), *Atlas Tethys Palaeoenvironmental Maps*. Explanatory Notes. Gauthier-Villars, Paris, 113–134; map 1/20.000.000 by Fourcade E., Azema J., Cecca F., Dercourt J., Vrielynck B., Bellion Y., Sandulescu M. and Ricou L. E. Late Tithonian Paleoenvironments. BEICIP-FRANLAB, Rueil-Malmaison.

Fourcade E. and Michaud F. (1987) L'ouverture de l'Atlantique et son influence sur les peuplements des grands foraminifères des plates-formes péri-océaniques au Mésozoïque. *Geodinamica Acta*, **1**, 247–262.

Gallagher W. B. (1993) The Cretaceous/Tertiary mass extinction event in the northern Atlantic coastal plain. *The Mosasaur*, **5**, 75–154.

Garagna S., Zuccotti M., Redi C. A. and Capanna E. (1997) Trapping speciation. *Nature*, **390**, 241–242.

Gardner R. N. and Campbell H. J. (1997) The bivalve genus *Grammatodon* from the Middle Jurassic of the Catlins district, South Otago, New Zealand. *New Zealand Journal of Geology and Geophysics*, **40**, 487–498.

Gaston K. J. (2000) Global patterns in biodiversity. *Nature*, **405**, 220–227.

Grande L. (1985) The use of paleontology in systematics and biogeography, and a time control refinement for historical biogeography. *Paleobiology*, 234–243.

Grant P. R. (1998) Competition exposed by knight? *Nature*, **396**, 216–217.

Grigg R. W. and Hey R. (1992) Palaeoceanography of the Tropical Eastern Pacific Ocean. *Science*, **255**, 172–178.

Hallam A. (1973) *Atlas of Palaeobiogeography*. Elsevier, Amsterdam.

Hallam A. (1974) Changing patterns of provinciality and diversity of fossil animals in relation to plate tectonics. *Journal of Biogeography*, **1**, 213–225.

Hallam A. (1975) *Jurassic Environments*. Cambridge, Earth Sci. Ser., Cambridge University Press.

Hallam A. (1977) Jurassic bivalve biogeography. *Paleobiology*, **3**, 58–73.

Hallam A. (1983) Early and mid Jurassic molluscan biogeography and the establishment of the Central Atlantic Seaway. *Palaeogeogr. Palaeoclimat. Palaeoecol.*, **43**, 181–193.

Hallam A. (1986) Evidence of displaced terranes from Permian to Jurassic faunas around the Pacific margins. *Jour. Geol. Soc. London*, **143**, 209–216.

Hallam A. (1987) Radiations and extinctions in relation to environmental change in the marine Lower Jurassic of northwest Europe. *Paleobiology*, **13**, 152–168.

Hallam A. (1988) A re-evaluation of Jurassic eustasy in the light of new data and the revised Exxon curve. *S. E. P. M., spec. publ.*, **42**, 261–273.

Hallam A. (1994) *An Outline of Phanerozoic Biogeography*. Oxford University Press.

Hallam A., Perez E. and Biro L. (1986) Facies analysis of the Lo Valdes Formation (Tithonian–Hauterivian) of the High Cordillera of central Chile and the palaeogeographic evolution of the Andean Basin. *Geological Magazine*, **123**, 425–435.

Hallam A. and Wignall P. (1997) *Mass Extinctions and Their Aftermath*. Oxford University Press.

Hallam A. and Wignall P. (1999) Mass extinctions and sea-level changes. *Earth-Science Reviews*, **48**, 217–250.

Hansen T. A. (1980) Influence of larval dispersal and geographic distribution on species longevity in neogastropods. *Paleobiology*, **6**, 193–207.

Hanski I. and Gilpin M. (1991) Metapopulation dynamics: brief history and conceptual domain. *Biol. Journ. Linnean Soc.*, **42**, 3–16.

Hantzpergue P. (1989) Les Ammonites kimméridgiennes du haut-fond d'Europe occidentale. Biochronologie, Systématique, Evolution, Paléobiogéographie. *Cahiers de Paléontologie*, 1–428, Paris.

Haq B. U., Hardenbold J. and Vail P. R. (1988) Mesozoic and Cenozoic chronostratigraphy and cycles of sea-level change. *S. E. P. M., spec. publ.*, **42**, 71–108.

Harrison P. L. and Wallace C. C. (1990) Reproduction, dispersal and recruitment of scleractinian corals. In: Dubinsky Z. (Ed.), *Coral Reefs*. Ecosystems of the World, **25**, 133–207, Elsevier, Amsterdam.

Hazel J. E. (1970) Binary coefficients and clustering in biostratigraphy. *Geol. Soc. Am. Bull.*, **81**, 3237–3252.

Henderson R. A. and Heron M. L. (1977) A probabilistic method of paleobiogeographic analysis. *Lethaia*, **10**, 1–15.

Hennig W. (1966) *Phylogenetic Systematics*. University of Illinois Press, Urbana, IL.

Hoedemaeker Ph. J. (1990) The Neocomian boundaries of the Tethyan Realm based on the distribution of ammonites. *Cretaceous Research*, **11**, 331–342.

Hoedemaeker Ph. J. (1995) Ammonite evidence for long-term sea-level fluctuations between the 2nd and 3rd order in the lowest Cretaceous. *Cretaceous Research*, **16**, 230–241.

Hoffmann A. and Szubzda-Studencka B. (1982) Bivalve species duration and ecologic characteristics in the Badenian (Miocene) marine sandy facies of Poland. *N. Jb. Geol. Paläont. Abh.*, **163**, 193–207.

Holder M. T., Erdmann M. V., Wilcox T. P., Caldwell R. and Hillis D. M. (1999) Two living species of coelacanths? *Proc. Nat. Acad. Sciences*, **96** (22), 12616–12620.

House M. R. (1985) Correlation of mid-Palaeozoic ammonoid evolutionary events with global sedimentary perturbations. *Nature*, **313**, 17–22.

Hughes C. P. (1973) Analysis of past faunal distributions. In: Tarling D. H. and Runcorn S. K. (Eds), *Implications of Continental Drift to the Earth Sciences*, 221–230. Academic Press, London.

Hughes T. P., Baird A. H., Dinsdale E. A., Moltschaniwskyj N. A., Pratchett M. S., Tanner J. E. and Willis B. L. (1999) Patterns of recruitment and abundance of corals along the Great Barrier Reef. *Nature*, **397**, 59–63.

Humphries C. J. and Parenti L. (1986) *Cladistic Biogeography*. Oxford Monographs on Biogeography. Clarendon Press, Oxford.

Ivany L. C., Patterson W. P. and Lohmann K. C. (2000) Cooler winters as a possible cause of mass extinctions at the Eocene/Oligocene boundary. *Nature*, **407**, 887–890.

Jablonski D. (1980) Apparent versus real biotic effects of transgression and regression. *Paleobiology*, **6**, 397–407.

Jablonski D. (1985) Marine regressions and mass extinctions: a test using the modern biota. In: Valentine J. W. (Ed.), *Phanerozoic Diversity Patterns*, 335–354. Princeton University Press, Princeton, NJ.

Jablonski D. (1986a) Causes and consequences of mass extinctions: a comparative approach. In: Elliott D. K. (Ed.), *Dynamics of Extinction*, 183–229. Wiley-Interscience Publishers, New York.

Jablonski D. (1986b) Background and mass extinctions: the alternation of macroevolutionary regimes. *Science*, **231**, 129–133.

Jablonski D. (1986c) Larval ecology and macroevolution in marine invertebrates. *Bull. Mar. Sci.*, **39**, 565–587.

Jablonski D. (1993) The tropics as a source of evolutionary novelty through geological time. *Nature*, **364**, 142–144.

Jablonski D. (1998) Geographic variation in the moluscan recovery from the end-Cretaceous extinctions. *Science*, **279**, 1327–1330.

Jablonski D., Flessa K. W. and Valentine J. W. (1985) Biogeography and paleobiology. *Paleobiology*, **11**, 75–90.

Jablonski D. and Lutz R. A. (1980) Larval shell morphology: ecological and paleontological applications. In: Rhoads D. C. and Lutz R. A. (Eds), *Skeletal Growth of Aquatic Organisms*, 323–377. Plenum, New York.

Jablonski D. and Lutz R. A. (1983) Larval ecology of marine benthic invertebrates: paleobiological implications. *Biological Revue*, **58**, 21–89.

Jackson J. B. C., Budd A. F. and Pandolfi J. M. (1996) The shifting balance of natural communities. In Jablonski D., Erwin D. H. and Lipps J. H. (Eds), *Evolutionary Paleobiology*, 89–122. The University of Chicago Press, Chicago and London.

Jakobs G. K. (1995) New occurrences of *Leukadiella* and *Paroniceras* (Ammonoidea) from the Toarcian (Lower Jurassic) of the Canadian Cordillera. *J. Paleont.*, **69**, 89–98.

Jeffery C. H. (2001) Heart urchins at the Cretaceous/Tertiary boundary: a tale of two clades. *Paleobiology*, **27**, 140–158.

Kauffman E. G. (1973) Cretaceous Bivalvia. In: Hallam A. (Ed.), *Atlas of Paleobiogeography*, 353–382. Elsevier, New York.

Keller G. (1983) Paleoclimatic analyses of middle Eocene through Oligocene planktic foraminiferal faunas. *Palaeogeogr. Palaeoclimatol. Palaeoecol.*, **43**, 73–94.

Kemper E., Rawson P. F. and Thieuloy J.-P. (1981) Ammonites of Tethyan ancestry in the early Lower Cretaceous of north-west Europe. *Palaeontology*, **24**, 251–311.

Koch C. F. (1987) Prediction of sample size effects on the measured temporal and geographic distribution patterns of species. *Paleobiology*, **13**, 100–107.

Kondrashov A. S. and Kondrashov F. A. (1999) Interactions among quantitative traits in the course of sympatric speciation. *Nature*, **400**, 351–354.

Koutsoukos E. A. M. (1992) Late Aptian to Maastrichtian foraminiferal biogeography and palaeoceanography of the Sergipe Basin, Brazil. *Palaeogeogr. Palaeoclimatol. Palaeoecol.* **92**, 295–324.

Kutek J. and Zeiss A. (1974) Tithonian–Volgian ammonites from Brzostowka near Tomaszow Mazowiecki, Central Poland. *Acta Geol. Pol.*, **24**, 505–542.

Landman N., Tanabe K. and Davis R. A. (Eds) (1996) *Ammonoid Paleobiology*. Topics in Geobiology, **13**, Plenum Press, New York.

Laurin B. (1992) Découverte d'un squelette de soutien du lophophore de type 'crura' chez un brachiopode inarticulé: description de *Neoancistrocrania norfolki* gen. sp. nov. (Craniidae). *C. R. Acad. Sciences Paris*, sér. III, **314**, 343–350.

Lethiers F. (1998) *Evolution de la biosphère et événements géologiques*. Gordon & Breach Science Publishers, Reading.

Lethiers F. and Crasquin-Soleau S. (1995) Distribution des ostracodes et paléocourantologie au Carbonifère terminal–Permien. *Geobios*, M. S., **18**, 257–272.

Lieberman B. S. (1997) Early Cambrian paleogeography and tectonic history: a biogeographic approach. *Geology*, **25**, 1039–1042.

Lieberman B. S. (2000) *Paleobiogeography: Using Fossils to Study Global Change, Plate Tectonics, and Evolution*. Topics in geobiology, **16**, Kluwer Academic/Plenum Publishers, New York.

Lieberman B. S. and Eldredge N. (1996) Trilobite biogeography in the Middle Devonian: geological processes and analytical methods. *Paleobiology*, **22**, 66–79.

Lutz A. and Jablonski D. (1982) L'utilisation de la morphologie des coquilles larvaires et postlarvaires de Bivalves pour les études biogéographiques et paléobiogéographiques. *Bull. Soc. Géol. France*, sér. 7, **24** (5–6), 901–906.

MacArthur R. H. and Wilson E. O. (1963) An equilibrium theory of insular zoogeography. *Evolution*, **17**, 373–387.

MacArthur R. H. and Wilson E. O. (1967) *The Theory of Island Biogeography*. Princeton University Press, Princeton, NJ.

MacLeod N. and Keller G. (1994) Comparative biogeographic analysis of planktonic foraminiferal survivorship across the Cretaceous/Tertiary (K/T) boundary. *Paleobiology*, **20**, 143–177.

McCoy E. D. and Heck K. L. (1976) Biogeography of corals, seagrasses and mangroves: an alternative to the center of origin concept. *Syst. Zool.*, **25**, 201–210.

McCoy E. D. and Heck K. L. (1983) Centers of origin revisited. *Paleobiology*, **9**, 17–19.

McGowran B., Li Q., Cann J., Padley D., McKirdy D. M. and Shafik S. (1997) Biogeographic impact of the Leeuwin Current in southern Australia since the late middle Eocene. *Palaeogeogr. Palaeoclimat. Palaeoecol.*, **136**, 19–40.

McKenna M. C. (1973) Sweepstakes, filters, corridors, Noah's Arks and beached viking funeral ships in paleogeography. In: Tarling, D. N. and Runcorn S. K (Eds), *Implications of Continental Drift to the Earth Sciences*, 295–308. Academic Press, London.

McKerrow W. S. and Cocks L. R. M. (1986) Oceans, island arcs and olistostromes: the use of fossils in distinguishing sutures, terranes and environments around the Iapetus Ocean. *Journ. Geol. Soc., London*, **143**, 185–191.

McKerrow W. S. and Scotese C. R. (1990) *Palaeozoic Palaeogeography and Biogeography. Mem. Geol. Soc. London*, **12**.

Marcoux J., Baud A., Ricou L.-E., Gaetani M., Krystyn L., Bellion Y., Guiraud R., Moreau C., Besse J., Gallet Y., Jaillard E. and Théveniaut H. (1993) Late Anisian (237-234 Ma). In: Dercourt J., Ricou L. E. and Vrielynck, B. (Eds), *Atlas Tethys Palaeoenvironmental Maps*. Explanatory Notes. Gauthier-Villars, Paris, 21–34.

Margules C. R. and Pressey R. L. (2000) Systematic conservation planning. *Nature*, **405**, 243–253.

Masse J.-P. (1992) The Lower Cretaceous Mesogean benthic ecosystems: palaeoecologic aspects and palaeobiogeographic implications. *Palaeogeogr. Palaeoclim. Palaeoecol.*, **91**, 331–345.

Matthew W. D. (1915) Climate and evolution. *Ann. N. Y. Acad. Sci.*, **24**, 171–318.

Matyja B. and Wierzbowski A. (2000) Biological response of ammonites to changing environmental conditions: an example of Boreal Amoeboceras invasions into Submediterranean Province during Late Oxfordian. *Acta Geologica Polonica*, **50**, 45–54.

Mayr E. (1940) Speciation phenomena in birds. *Amer. Nat.*, **74**, 249–278.

Mayr E. (1942) *Systematics and the Origin of Species*. Dover Press, New York.

Mayr E. (1963) *Animal species and evolution*. The Belknap Press of Harvard University, Cambridge, MA.

Meister C. and Stampfli G. (2000) Les ammonites du Lias moyen (Pliensbachien) de la Néotéthys et de ses confins; compositions fauniques, affinités paléogéographiques et biodiversité. *Revue Paléobiol.*, **19**, 227–292.

Middlemiss F. A., Rawson P. F. and Newall G. (1971) *Faunal Provinces in Space and Time. Geological Journal*, Spec. Issue, **4**, Seel House Press, Liverpool.

Miller M. M. and Wright J. E. (1987) Paleogeographic implications of Permian Tethyan corals from the Klamath Mountains, California. *Geology*, **15**, 266–269.

Mizutani S. and Kojima S. (1992) Mesozoic radiolarian biostratigraphy of Japan and collage tectonics along the eastern continental margin of Asia. *Palaeogeogr. Palaeoclimat. Palaeoecol.*, **96**, 3–22.

Monge-Najera J. (1995) Phylogeny, biogeography and reproductive trends in the Onychophora. *Zool. Journ. Linnean Soc.*, **114**, 21–60.

Monge-Najera J. (1996) Jurassic–Pliocene biogeography: testing a model with velvet worm (Onychophora) vicariance. *Rev. Biol. Trop.*, **44**, 159–175.

Monger J. W. H. and Ross C. A. (1971) Distribution of fusulinaceans in the western Canadian Cordillera. *Canadian Journal of Earth Sciences*, **8**, 259–278.

Monnet C. and Bucher H. (1999) Biochronologie quantitative (associations unitaires) des faunes d'ammonites du Cénomanien du Sud-Est de la France. *Bull. Soc. Géol. France*, **170**, 599–610.

Montaggioni L. (2000) Postglacial reef growth. *C. R. Acad. Sciences Paris*, sér. II, **331**, 319–330.

Morris N. J. and Skelton P. W. (1995) Late Campanian–Maastrichtian rudists from the United Arab Emirates–Oman border region. *Bull. Nat. Hist. Mus. London* (Geol.), **51**, 277–305.

Mueller A. H. (1970) Über den Sexualdimorphismus regulärer Echinoidea (Echinodermata) der Oberkreide. Monats. *Deutsche Akad. Wissensch. Berlin*, **12**, 923–935.

Mutterlose J. (1986) Upper Jurassic belemnites from the Orville Coast, Western Antarctica, and their palaeobiogeographic significance. *British Antarctic Survey Bulletin*, **70**, 1–22.

Nelson G. (1973) Comments on Leon Croizat's biogeography. *Syst. Zool.*, **22**, 312–320.

Nelson G. (1983) Vicariance and cladistics: historical perspectives with implications for the future. In: Sims R. W., Price J.-H. and Whalley P. E. S. (Eds), *Evolution, Time and Space: the Emergence of the Biosphere*, 469–492. *Syst. Ass.*, spec. vol., **23**, Academic Press, London and New York.

Nelson G. and Platnick N. I. (1980) A vicariance approach to historical biogeography. *BioScience*, **30**, 339–343.

Nelson G. and Platnick N. I. (1981) *Systematics and Biogeography – cladistics and vicariance*. Columbia University Press, New York.

Néraudeau D. and Mathey B. (2000) Biogeography and diversity of South Atlantic Cretaceous echinoids: implications for circulation patterns. *Palaeogeogr. Palaeoclimat. Palaeoecol.*, **156**, 71–88.

Neumayr M. (1883) Über klimatische Zonen wahrend der Jura-und Kreidezeit. *Denkschr. Akad. Wiss. Wien*, **47**, 277–310.

Newell N. D. (1971) An outline history of tropical organic reefs. *American Museum Novitiates*, **2465**, 1–37.

Newton C. R. (1987) Biogeographic complexity in Triassic bivalves of the Wallowa terrane, northwestern United States: oceanic islands, not continents, provide the best analogies. *Geology*, **15**, 1126–1129.

Newton C. R. (1988) Significance of 'Tethyan' fossils in the American Cordillera. *Science*, **242**, 385–391.

Newton C. R. (1990) Palaeobiogeography of the ancient Pacific, response. *Science*, **249**, 681–683.

Norris R. D. (2000) Pelagic species diversity, biogeography and evolution. In: Erwin D. H. and Wing S. L. (Eds), *Deep Time, Paleobiology's Perspective*, 236–258. *Paleobiology*, suppl. to vol. **26** (4).

O'Brien E. M. (1993) Climatic gradients in woody plant species richness: towards an explanation based on an analysis of southern Africa's woody flora. *Journal of Biogeography*, **20**, 181–198.

O'Brien E. M. (1998) Water–energy dynamics, climate, and prediction of woody plant species richness: an interim general model. *Journal of Biogeography*, **25**, 379–398.

Ockelmann K. W. (1965) Developmental types in marine bivalves and their distribution along the Atlantic coast of Europe. In: Cox L. R. and Peake J. F. (Eds), *Proceedings of the First European Malacological Congress*, London, 1962, 25–35. Conchological Society of Great Britain and Ireland and the Malacological Society of London.

O'Dogherty L., Sandoval J. and Vera J. A. (2000) Ammonite faunal turnover tracing sea-level changes during the Jurassic (Betic Cordillera, southern Spain). *Journ. Geol. Soc., London*, **157**, 723–736.

Oloriz F. (1986) Paleogeography and Ammonites in the Upper Jurassic. Outlines for a pattern. In: Pallini G. (Ed.), *Fossili, Evoluzione, Ambiente*, 1–9, Atti I° Convegno Pergola, 1984.

Oloriz F. (1990) Ammonite phenotypes and Ammonite distributions. Notes and comments. In: Pallini G., Cecca F., Cresta S. and Santantonio M. (Eds), *Fossili, Evoluzione, Ambiente*, 417–426, Atti II° Convegno Pergola, 1987.

Oloriz F., Marques B. and Rodriguez-Tovar F. J. (1991) Eustatism and faunal associations. Examples from the South Iberian Margin during the Late Jurassic (Oxfordian–Kimmeridgian). *Eclogae Geol. Helv.*, **84**, 83–106.

Oloriz F., Rodriguez-Tovar F. J., Marques B. and Caracuel J. (1993) Ecostratigraphy and sequence stratigraphy in high frequency sea level fluctuations: examples from Jurassic macroinvertebrate assemblages. *Palaeogeogr. Palaeoclimatol. Palaeoecol.*, **101**, 131–145.

Page R. D. M. (1987) Graphs and generalized tracks: quantifying Croizat's Panbiogeography. *Syst. Zool.*, **36**, 1–17.

Pandolfi J. M. (1992) A review of the tectonic history of New Guinea and its significance for marine biogeography. *Proceedings of the Seventh Int. Coral Reef Symposium*, Guam, **2**, 718–728.

Paris F. and Robardet M. (1990) Early Paleozoic palaeobiogeography of the Variscan regions. *Tectonophysics*, **177**, 193–213.

Parker W. C., Feldman A. and Arnold A. J. (1999) Paleobiogeographic patterns in the morphologic diversification of the Neogene planktonic foraminifera. *Palaeogeogr. Palaeoclimatol. Palaeoecol.*, **152**, 1–14.

Parrish J. T. (1992) Jurassic climate and oceanography of the Pacific region. In: Westermann G. E. G. (Ed.), *The Jurassic of the Circum-Pacific*, 365–379. Cambridge University Press.

Patterson C. (1981) Methods of palaeobiogeography. In: Nelson G. and Rosen D. E. (Eds), *Vicariance Biogeography: a Critique*, 446–489. Columbia University Press, New York.

Patterson C. (1983) Aims and methods in biogeography. In: Sims R. W., Price J.-H. and Whalley P. E. S. (Eds), *Evolution, Time and Space: the Emergence of the Biosphere*, 1–28. *Syst. Ass.*, spec. vol., **23**, Academic Press, London and New York.

Philip J. (1982) Paléobiogéographie des Rudistes et géodynamique des marges mésogéennes au Crétacé supérieur. *Bull. Soc. Géol. France*, sér. 7, **24** (5–6), 995–1006.

Philip J. (1985) Sur les relations des marges téthysiennes au Campanien et au Maastrichtien déduites de la distribution des Rudistes. *Bull. Soc. Géol. France*, sér. 8, **1** (5), 723–731.

Philip J. (1998) Biostratigraphie et paléobiogéographie des rudistes: évolution des concepts et progrès récents. *Bull. Soc. Géol. France*, **169** (5), 689–708.

Philip J. and Platel J.-P. (1994) *Praetorreites*, nouveau genre de rudiste du Campanien d'Oman. *Geobios*, **27**, 303–319.

Philip J. and Platel J.-P. (1998a) *Monopilarites*, un nouveau genre de rudiste du Campanien supérieur du Massif du Haushi-Huqf (Sultanat d'Oman). *Geobios, M. S.* **22**, 269–282.

Philip J. and Platel J.-P. (1998b) Nouvelle espèce de *Torreites* du Campanien du Haushi-Huqf, Sultanat d'Oman. *Geobios, M. S.* **22**, 283–294.

Piccoli G. and Sartori S. (1994) Paleobiogeography of the Tethys: a new method. *Boll. Soc. Paleont. Ital.*, spec. vol. **2**, 227–242.

Piccoli G., Sartori S. and Franchino A. (1986) Mathematical model of the migration of Cenozoic benthic molluscs in the Tethyan belt. *Mem. Sci. Geol.*, **38**, 207–244.

Pielou E. C. (1966) Species diversity and pattern diversity in the study of ecological succession. *Journ. Theor. Biology*, **10**, 370–383.

Platnick N. I. (1981) The progression rule or progress beyond rules in biogeography. In: Nelson G. and Rosen D. E. (Eds), *Vicariance Biogeography: a Critique*, 144–150. Columbia University Press, New York.

Platnick N. I. and Nelson G. (1978) A method of analysis for historical biogeography. *Syst. Zool.*, **27**, 1–16.

Pouyaud L., Wirjoatmodjo S., Rachmatika I., Tjakrawidjaja A., Hadiaty R. and Hadie W. (1999) A new species of coelacanth. *C. R. Acad. Sciences Paris*, sér. III, **322**, 261–267.

Preston F. W. (1962) The canonical distribution of commonness and rarity. *Ecology*, **43**, 185–215, 410–432.

Purvis A. and Hector A. (2000) Getting the measure of biodiversity. *Nature*, **405**, 212–219.

Raffi S. and Taviani M. (1984) An outline of the late Tertiary to Recent history of the Mediterranean deep-sea mollusk fauna. *Ann. Géol. Pays Helléniques*, **32**, 257–261.

Rapoport E. H. (1975) *Areografia: estrategias geograficas de las especies*. Fondo de Cultura Economia, Mexico.

Rawson P. F. (1973) Lower Cretaceous (Ryazanian–Barremian) marine connections and cephalopod migrations between the Tethyan and Boreal Realms. In: Casey R. and Rawson P. F. (Eds), *The Boreal Lower Cretaceous*, 131–144. *Geol. J.*, spec. iss., **5**, Seel House Press, Liverpool.

Rawson P. F. (1993) The influence of sea-level changes on the migration and evolution of early Cretaceous (pre-Aptian) ammonites. In: House M. R. (Ed.), *The Ammonoidea: Environment, Ecology and Evolutionary Change*, 227–242. *Syst. Ass.*, spec. vol. **47**, Clarendon Press, Oxford.

Remane J. (1998) Les calpionelles: possibilités biostratigraphiques et limitations paléobiogéographiques. *Bull. Soc. Géol. France*, **169** (6), 829–839.

Riccardi A. C. (1991) Jurassic and Cretaceous marine connections between the Southeast Pacific and Tethys. *Palaeogeogr. Palaeoclim. Palaeoecol.*, **87**, 155–189.

Richer de Forges B., Koslow J. A. and Poore G. C. B. (2000) Diversity and endemism of the benthic seamount fauna in the southwest Pacific. *Nature*, **405**, 944–947.

Ricou L.-E. (1994) Tethys reconstructed: plates, continental fragments and their boundaries since 260 Ma from Central America to South-eastern Asia. *Geodinamica Acta*, **7**, 169–218.

Ron S. (2000) Biogeographic area relationships of lowland Neotropical rainforest based on raw distributions of vertebrate groups. *Biol. Journ. Linnean Society*, **71**, 379–402.

Rosen B. R. (1983) Reef island staging posts and Noah's arks. *Reef Encounter*, **1**, 5–6.

Rosen B. R. (1984) Reef coral biogeography and climate through the Late Cainozoic: just islands in the sun or a critical pattern of islands? In: Brenchley P. (Ed.), *Fossils and Climate*, 201–262. *Geol. J.*, spec. issue II, John Wiley and Sons, Chichester.

Rosen B. R. (1985) Long-term geographical controls on regional diversity [lecture transcript summary]. *J. Open Univ. Geol. Soc.*, **6**, 25–30.

Rosen B. R. (1988a) Biogeographic patterns: a perceptual overview. In: Myers A. A. and Giller P. S. (Eds), *Analytical Biogeography: an Integrated Approach to the Study of Animal and Plant Distributions*, 23–55. Chapman & Hall, London.

Rosen B. R. (1988b) From fossils to earth history: applied historical biogeography. In: Myers A. A. and Giller P. S. (Eds), *Analytical Biogeography: an Integrated Approach to the Study of Animal and Plant Distributions*, 437–481. Chapman & Hall, London.

Rosen B. R. (1988c) Progress, problems and patterns in the biogeography of reef corals and other tropical marine organisms. *Helgoländer Meeresunters*, **42**, 269–301.

Rosen B. R. (1992) Empiricism and the biogeographical black box: concepts and methods in marine palaeobiogeography. *Palaeogeogr. Palaeoclimatol. Palaeoecol.*, **92**, 171–205.

Rosen B. R. (1995) Neobiogeography versus palaeobiogeography. *Boll. Soc. Paleont. Ital.*, spec. vol. **2**, 291–303.

Rosen B. R. and Smith A. B. (1988) Tectonics from fossils? Analysis of reef-coral and sea-urchin distributions from late Cretaceous to Recent, using a new method. In: Audley-Charles M. G. and Hallam A. (Eds), *Gondwana and Tethys*, 275–306. Geological Society, spec. publ. 37, London.

Rosen B. R. and Turnsek D. (1989) Extinction patterns and biogeography of scleractinian corals across the Cretaceous/Tertiary boundary. *Mem. Ass. Australas. Palaeontols*, **8**, 355–370.

Rosen D. E. (1979) Fishes from the uplands and intermontane basins of Guatemala: revisionary studies and comparative geography. *Bull. Amer. Mus. Nat. Hist.*, **162**, 267–376.

Rosenzweig M. L. (1977) Does the fossil record provide for natural experiments? *Palaeobiology*, **3**, 322–324.

Rosenzweig M. L. (1995) *Species Diversity in Space and Time.* Cambridge University Press.

Ross C. A. and Ross J. R. P. (1985) Carboniferous and Early Permian biogeography. *Geology*, **13**, 27–30.

Rosso A. and Di Geronimo I. (1998) Deep-sea Pleistocene Bryozoa of Southern Italy. *Geobios*, **30**, 303–317.

Rotondo G. M., Springer V. G., Scott G. A. J. and Schlanger S. O. (1981) Plate movement and island integration – a possible mechanism in the formation of endemic biotas, with special reference to the Hawaiian Islands. *Syst. Zool.*, **30**, 12–21.

Roux M. (1982) De la biogéographie historique des océans aux reconstitutions paléobiogéographiques: tendances et problèmes illustrés par des exemples pris chez les Echinodermes bathyaux et abyssaux. *Bull. Soc. Géol. France*, sér. 7, **24** (5–6), 907–916.

Roy K., Jablonski D. and Valentine J. W. (1996) Higher taxa in biodiversity studies: patterns from eastern Pacific marine molluscs. *Phil. Trans. R. Soc. London*, **B351**, 1605–1613.

Roy K., Jablonski D. and Valentine J. W. (2000) Dissecting latitudinal diversity gradients: functional groups and clades of marine bivalves. *Proc. R. Soc. London*, **B267**, 293–299.

Roy K., Jablonski D., Valentine J. W. and Rosenberg G. (1998) Marine latitudinal diversity gradients: tests of causal hypotheses. *Proc. Nat. Acad. Sci. USA*, **95**, 3699–3702.

Scheltema R. S. (1968) Dispersal of larvae by equatorial ocean currents and its importance to the zoogeography of shoal-water tropical species. *Nature*, **217**, 1159–1162.

Scheltema R. S. (1971) The dispersal of the larvae of shoal-water benthic invertebrate species over long distances by ocean currents. In: Crisp D. J. (Ed.), *Fourth European Marine Biology Symposium*, 7–28. Cambridge University Press.

Scheltema R. S. (1978) On the relationship between dispersal of pelagic veliger larvae and the evolution of marine prosobranch gastropods. In: Battaglia B. and Beardmore J. A. (Eds), *Marine Organisms: Genetics, Ecology and Evolution*, 303–322. Plenum Press, New York.

Scheltema R. S. (1979) Dispersal of pelagic larvae and the zoogeography of Tertiary marine benthic gastropods. In: Gray J. and Boucot A. (Eds), *Historical Biogeography, Plate Tectonics and the Changing Environment*, 391–397. Oregon State University Press, Corvallis, OR.

Schlanger S. O., Jenkyns H. C. and Premoli Silva I. (1981) Volcanism and vertical tectonics in the Pacific Basin related to global Cretaceous transgressions. *Earth Planet. Sci. Lett.*, **52**, 435–449.

Schneider C. E. and Kennett J. P. (1999) Segregation and speciation in the Neogene planktonic foraminiferal clade *Globoconella. Paleobiology*, **25**, 383–395.

Schopf T. J. M. (1979) The role of biogeographic provinces in regulating marine faunal diversity through geologic time. In: Gray J. and Boucot A. (Eds), *Historical Biogeography, Plate Tectonics and the Changing Environment*, 449–457. Oregon State University Press, Corvallis, OR.

Sepkoski J. J. Jr. (1988) Alpha, beta or gamma: where does all the diversity go? *Paleobiology*, **14**, 221–234.

Servais T. and Mette W. (2000) The *messaoudensis-trifidum* acritarch assemblage (Ordovician: late Temadoc–early Arenig) of the Barriga Shale Formation, Sierra Morena (SW-Spain). *Rev. Palaeobot. Palynol.*, **113**, 145–163.

Sey I. I., Repin S., Kalacheva E. D., Okuneva T. M., Paraketsov K. V. and Polubotko I. V. (1992) Eastern Russia. In: Westermann G. E. G. (Ed.), *The Jurassic of the Circum-Pacific*, 225–245. Cambridge University Press.

Shi G. R. (1993) Multivariate data analysis in palaeoecology and palaeobiogeography – a review. *Palaeogeogr. Palaeoclimat. Palaeoecol.*, **105**, 199–234.

Shuto T. (1974) Larval ecology of prosobranch gastropods and its bearing on biogeography and paleontology. *Lethaia*, **7**, 239–256.

Simpson G. G. (1940) Mammals and land bridges. *J. Washington Acad. Sci.*, **30**, 137–163.

Simpson G. G. (1950) History of the fauna of Latin America. *Amer. J. Sci.*, **38**, 261–289.

Simpson G. G. (1960) Notes on the measurement of faunal resemblance. *Amer. J. Sci.*, **258-A**, 300–311.

Simpson G. G. (1962) *Evolution and Geography: an Essay on Historical Biogeography with Special Reference to Mammals*. 3rd edn. Condon Lectures, Oregon State System of Higher Education, Eugene, OR.

Skelton P. W. (1988) The trans-Pacific spread of equatorial shallow-marine benthos in the Cretaceous. In: Audley-Charles M. G. and Hallam A. (Eds), *Gondwana and Tethys*, 247–253. Geological Society, spec. publ. 37, London.

Skelton P. W., Crame J. A., Morris N. J. and Harper E. M. (1990) Adaptive divergence and taxonomic radiation in post-Palaeozoic bivalves. In: Taylor P. D. and Larwood G. P. (Eds), *Major Evolutionary Radiations*, 91–117. Syst. Ass., spec. vol. **42**, Clarendon Press, Oxford.

Skelton P. W. and Wright V. P. (1987) A Caribbean rudist bivalve in Oman: island-hopping in the Late Cretaceous. *Palaeontology*, **30**, 505–529.

Smith A. B. (1988) Late Palaeozoic biogeography of East Asia and palaeontological constraints on plate tectonic reconstructions. *Phil. Trans. R. Soc. London*, **A236**, 189–227.

Smith A. B. (1992) Echinoid distribution in the Cenomanian: an analytical study in biogeography. *Palaeogeogr. Palaeoclimatol. Palaeoecol.*, **92**, 263–276.

Smith P. L. (1983) The Pliensbachian ammonite Dayiceras dayiceroides and Early Jurassic paleogeography. *Canadian Journ. of Earth Sciences*, **20**, 86–91.

Smith P. L. and Tipper H. W. (1986) Plates tectonics and palaeobiogeography: early Jurassic (Pliensbachian) endemism and diversity. *Palaios*, **1**, 399–412.

Smith P. L. and Westermann G. E. G. (1990) Paleobiogeography of the Ancient Pacific. *Science*, **249**, 680.

Stanley G. D. Jr and Yancey T. E. (1990) Paleobiogeography of the Ancient Pacific. *Science*, **249**, 680–681.

Stebbins G. L. (1974) *Flowering Plants: Evolution above the Species Level.* Harvard University Press, Cambridge, MA.

Stehli F. G., McAlester A. L. and Helsley C. E. (1967) Taxonomic diversity of recent bivalves and some implications for geology. *Bull. Geol. Soc. America*, **78**, 455–466.

Stehli F. G. and Wells J. W. (1971) Diversity and age patterns in hermatypic corals. *Syst. Zool.*, **20**, 115–126.

Steininger F. F. and Rögl F. (1984) Paleogeography and palinspastic reconstruction of the Neogene of the Mediterranean and Paratethys. In: Dixon J. E. and Robertson A. H. F. (Eds), *The Geological Evolution of the Eastern Mediterranean*, 659–668. Blackwell Scientific Publications, Oxford.

Stevens C. H. (1983) Corals from a dismembered late Paleozoic paleo-Pacific plateau. *Geology*, **11**, 603–606.

Stevens G. C. (1989) The latitudinal gradient in geographical ranges: how so many species coexist in the tropics. *Amer. Nat.*, **133**, 240–256.

Strathmann R. R. (1978) The evolution and loss of feeding larval stages of marine invertebrates. *Evolution*, **32**, 894–906.

Swofford D. (1990) Phylogenetic analysis using parsimony (PAUP), version 3. Illinois State University Survey.

Sylvester-Bradley P. C. (1971) Dynamic factors in animal palaeogeography. In: Middlemiss F. A., Rawson P. F. and Newall G. (Eds), *Faunal Provinces in Space and Time*, 1–18. *Geological J.*, Spec. Issue, **4**, Seel House Press, Liverpool.

Taviani M. (1998) Stable tropics not so stable: climatically driven extinctions of reef-associated molluscan assemblages (Red Sea and western Indian Ocean; last interglaciation to present). *Spec. Publs Int. Ass. Sediment.*, **25**, 69–76.

Termier H. and Termier G. (1952) *Histoire géologique de la Biosphère*. Masson Édit., Paris.

Thorson G. (1950) Reproductive and larval ecology of marine bottom invertebrates. *Biological Reviews*, **25**, 1–45.

Thorson G. (1957) Bottom communities (sublittoral or shallow shelf). *Geol. Soc. Amer. Mem.*, **67**, 461–534.

Torsvik T. H., Tait J., Moralev V. M., McKerrow W. S., Sturt B. A. and Roberts D. (1995) Ordovician palaeogeography of Siberia and adjacent continents. *Journ. Geol. Soc. London*, **152**, 279–287.

Torsvik T. H. and Trench A. (1991) The Ordovician history of the Iapetus Ocean in Britain: new palaeomagnetic constraints. *Journ. Geol. Soc. London*, **148**, 423–426.

Tozer E. T. (1982) Marine Triassic faunas of North American: their significance for assessing plate and terrane movements. *Geol. Rundschau*, **71**, 1077–1104.

Tregenza T. and Butlin R. K. (1999) Speciation without isolation. *Nature*, **400**, 311–312.

Uhlig V. (1911) Die marinen Reiche des Jura und der Unterkreide. *Mitt. Geol. Ges. Wien*, **4**, 229–448.

Vacelet J., Cuif J.-P., Gautret P., Massot M., Richer de Forges B. and Zimbrowius H. (1992) Un Spongiaire Sphinctozoaire colonial apparenté aux constructeurs de récifs triasiques survivant dans le bathyal de Nouvelle Calédonie. *C. R. Acad. Sciences Paris*, sér. III, **314**, 379–385.

Vachard D., Flores De Dios A., Pantoja J., Buitron B. E., Arellano J. and Grajales M. (2000) Les Fusulines du Mexique, une revue biostratigraphique et paléogéographique. *Geobios*, **33**, 655–679.

Vachard D. and Fourcade E., with the collaboration of Romero J. E., Mendez J., Cosillo A., Alonzo M., Requena J., Azéma J. and Cros P. (1997) Foraminifères et Algues du Permien du Guatemala. *Geobios*, **30**, 745–784.

Valentine J. W. (1973) *Evolutionary Paleoecology of the marine biosphere.* Prentice Hall, Englewood Cliffs, NJ.

Valentine J. W. and Jablonski, D. (1982) Major determinants of the biogeographic pattern of the shallow-sea fauna. *Bull. Soc. Géol. France*, sér. 7, **24** (5–6), 893–899.

Valentine J. W. and Jablonski, D. (1983a) Larval adaptations and patterns of brachiopod diversity in space and time. *Evolution*, **37**, 1052–1061.

Valentine J. W. and Jablonski, D. (1983b) Speciation in the shallow sea: general patterns and biogeographic controls. In: Sims R. W., Price J.-H. and Whalley P. E. S. (Eds), *Evolution, Time and Space: the Emergence of the Biosphere*, 201–226. *Syst. Ass.*, spec. vol., **23**, Academic Press, London and New York.

Valentine J. W. and Jablonski D. (1986) Mass extinctions: sensitivity of marine larval types. *Proc. Nat. Acad. Sci. USA*, **83**, 6912–6914.

Valentine J. W. and Jablonski D. (1991) Biotic effects of sea level change: the Pleistocene test. *Journal of Geophysical Research*, **96**, B4, 6873–6878.

Valentine J. W. and Moores E. M. (1970) Plate-tectonic regulation of faunal diversity and sea-level: a model. *Nature*, **228**, 657–659.

van der Spoel S. (1983) Patterns in plankton distribution and the relation to speciation: the dawn of the pelagic biogeography. In: Sims R. W., Price J.-H. and Whalley P. E. S. (Eds), *Evolution, Time and Space: the Emergence of the Biosphere*, 291–334. *Syst. Ass.*, spec. vol. **23**, Academic Press, London and New York.

Vermeij G. J. (1991) Symposium on biotic interchange: an introduction. *Paleobiology*, **17**, 201.

Vermeij G. J. (1995) Economics, volcanoes, and Phanerozoic revolutions. *Paleobiology*, **21**, 125–152.

Veron J. E. N. (1995) *Corals in Space and Time. The Biogeography and Evolution of the Scleractinia*. University of New South Wales Press, Sydney.

Von Hillebrandt A. (1981) Kontinentalverschiebung und die paläozoogeographischen Beziehungen des südamerikanischen Lias. *Geologische Rundschau*, **70**, 570–582.

Waggoner B. (1999) Biogeographic analyses of the Ediacara biota: a conflict with paleotectonic reconstructions. *Paleobiology*, **25**, 440–458.

Wallace A. R. (1876) *The Geographical Distribution of Animals*. Macmillan, London.

Wallace A. R. (1880) *Island Life*. Macmillan, London.

Waller T. R. (1991) Evolutionary relationships among commercial scallops (Mollusca: Bivalvia: Pectinidae). In: Shumway S. E. (Ed.), *Scallops: Biology, Ecology and Aquaculture*, 1–73. Elsevier, Amsterdam.

Waller T. R. (1993) The evolution of '*Chlamys*' (Mollusca: Bivalvia: Pectinidae) in the tropical western Atlantic and eastern Pacific. *Amer. Malacol. Bull.*, **10**, 195–249.

Webb S. D. (1991) Ecogeography and the Great American interchange. *Paleobiology*, **17**, 266–280.

Weissert H. and Mohr H. (1996) Late Jurassic climate and its impact on carbon cycling. *Palaeogeogr. Palaeoclim. Palaeoecol.*, **122**, 27–43.

Westermann G. E. G. (Ed.) (1992) *The Jurassic of the Circum-Pacific*. Cambridge University Press, Cambridge.

Westermann G. E. G. (1993) Global bio-events in mid-Jurassic ammonites controlled by seaways. In: House M. R. (Ed.), *Ammonoidea: Environment, Ecology, and Evolutionary Change*, 187–226. Oxford University Press, Oxford.

Westermann G. E. G. (1996) Ammonoid life and habitat. In: Landman N., Tanabe K. and Davis R. A (Eds), *Ammonoid Paleobiology*, 607–707. Topics in Geobiology **13**, Plenum Press, New York.

Westermann G. E. G. (2000a) The marine faunal Realms of the Mesozoic: review and revision under the new guidelines for biogeographic classification and nomenclature. *Palaeogeogr. Palaeoclimat. Palaeoecol.*, **158**, 1–13.

Westermann G. E. G. (2000b) Biochore classification and nomenclature in paleobiogeography: an attempt at order. *Palaeogeogr. Palaeoclimat. Palaeoecol.*, **163**, 49–68.

Westrop S. R. (1991) Intercontinental variation in mass extinction patterns: influence of biogeographic structure. *Paleobiology*, **17**, 363–368.

Whatley R. C. (1988) Ostracoda and palaeobiogeography. In: De Deckker P., Colin J.-P. and Peypouquet J.-P. (Eds), *Ostracoda in the Earth Sciences*, 103–123. Elsevier, Amsterdam.

Whittaker R. H. (1972) Evolution and measurement of species diversity. *Taxon*, **21**, 213–251.

Whittaker R. H. (1977) Evolution of species diversity in land communities. *Evolutionary Biology*, **10**, 1–67.

Whittaker R. J. (1998) *Island Biogeography. Ecology, Evolution, and Conservation*. Oxford University Press.

Whittington H. B. and Hughes C. P. (1972) Ordovician geography and faunal provinces deduced from trilobite distribution. *Philos. Trans. R. Soc. London*, **B263**, 235–278.

Wignall P. B. and Benton M. J. (1999) Lazarus taxa and fossil abundance at times of biotic crisis. *Jour. Geol. Soc. London*, **156**, 453–456.

Wiley E. O. (1980) Phylogenetic systematics and vicariance biogeography. *Syst. Bot.*, **5**, 194–220.

Wiley E. O. (1988) Parsimony analysis and vicariance biogeography. *Syst. Zool.*, **37**, 271–290.

Williams A. (1973) Distribution of brachiopod assemblages in relation to Ordovician palaeogeography. *Spec. Pap. Palaeontol.*, **12**, 241–269.

Willis J. C. (1922) *Age and area*. Cambridge University Press, London.

Wilson J. T. (1966) Did the Atlantic close and then re-open? *Nature, London*, **211**, 676–681.

Wilson M. E. J. and Rosen B. R. (1998) Implications of paucity of corals in the Paleogene of SE Asia: plate tectonics or Centre of Origin? In: Hall R. and Holloway J. D. (Eds), *Biogeography and Geological Evolution of SE Asia*, 165–195. Backhuys Publishers, Leiden.

Wyatt A. R. (1987) Shallow water areas in space and time. *Journ. Geol. Soc. London*, **144**, 115–120.

Wyatt A. R. (1993) Phanerozoic shallow water diversity driven by changes in sea-level. *Geolog. Rundschau*, **82**, 203–211.

Wyatt A. R. (1995) Late Ordovician extinctions and sea-level change. *Journ. Geol. Soc. London*, **152**, 899–902.

Zakharov V. A., Kurushin N. I. and Pokhialainen V. P. (1996) Paleobiogeographic criteria of terrane geodynamics of northeastern Asia in Mesozoic. *Russian Geology and Geophysics*, **37**, 1–22.

Ziegler A. M., Bambach R. K., Parrish J. T., Barrett S. F., Gierlowski E. H., Parker W. C., Raymond A. and Sepkoski J. J. (1981) Paleozoic biogeography and climatology. In: Niklas K. J. (Ed.), *Paleobotany, Paleoecology, and Evolution*, 231–267. Praeger Publishers, New York.

Ziegler B. (1967) Ammoniten ökologie am Beispeil des Oberjura. *Geolog. Rundschau*, **56**, 439–464.

Ziegler B. (1974) Grenzen der Biostratigraphie im Jura und gedanken zur stratigraphischen Methodik. *Mém. B. R. G. M.*, **75** (1971), 35–67.

Zunino M. (1999) A new concept of theoretical biogeography: the 'nemophoront'. In: Sobti and Yadav, *Some Aspects of the Insight of Insect Biology*, 109–113, Punjab University Press.

Zunino M. and Zullini A. (1995) *Biogeografia. La Dimensione Spaziale dell'Evoluzione*. Casa Editrice Ambrosiana, Milan.

Index